Words of Science

By the same author

FICTION

Pebble in the Sky

I, Robot

The Stars Like Dust

Foundation

Foundation and Empire

The Currents of Space

Second Foundation

The Caves of Steel

The Martian Way and Other Stories

The End of Eternity

The Naked Sun

Earth is Room Enough

The Death Dealers

Nine Tomorrows

NON-FICTION

Biochemistry and Human Metabolism *

The Chemicals of Life

Races and People *

Chemistry and Human Health *

Inside the Atom

Building Blocks of the Universe

Only a Trillion

The World of Carbon

The World of Nitrogen

** in collaboration*

Words of Science

and the History behind Them

Isaac Asimov

Illustrated by William Barss

Houghton Mifflin Company Boston

To the women in my life

GERTRUDE

and

ROBYN

c 16 15 14 13 12 11 10

LIBRARY OF CONGRESS CATALOG CARD NUMBER: 59-5198
ISBN: 0-395-06571-2

PRINTED IN THE U.S.A.

Introduction

MOST PEOPLE have had at least a glancing blow from one or more of the sciences in school, and if there is one impression they come out with, it is that science is "hard."

In some ways, of course, it is, just as any other subject is "hard." It isn't easy to learn to be a good carpenter, or a good actor, or a good poker player. The trouble is, though, that most people are early convinced that there's something particularly hard about science beyond the ordinary hardness of other skills.

Why?

One reason is the scientific vocabulary. Entering the world of mathematics and science turns into a meeting with a whole realm of new words: words that look and sound odd; words that are long and hard to pronounce; words that the ordinary person never meets with in ordinary life. It is as though scientists were protecting their mysteries from the prying eyes of ordinary mortals by an enveloping shroud of forbidding syllables.

And yet quite the reverse is true. The scientific vocabulary is the bridge by which we enter the land, not the wall that keeps us out. This shows up at those times when the scientific vocabulary fails us. For instance, there *are* scientific concepts which are expressed in common, ordinary words. An example is the word *work*. To the scientist, work is motion against a resisting force. Thus, lifting a rock against the force of gravity is work; driving a nail into resisting wood is work.

However, holding a piece of luggage six inches above the ground and motionless is *not* work; banging a nail against steel which it does not penetrate is *not* work. In ordinary English, however, work is anything you don't particularly want to do which makes you physically tired, so that holding luggage and banging uselessly at nails *is* work.

Naturally, students starting their physics course have trouble with the word *work*. It convinces some that scientists are a little queer. How much better off physicists would have been to have made up their own word for work.

In most cases, words *are* made up, and the words are usually taken from the Latin and Greek. This is a practice that began in the most natural way possible. In the first place, in the dawn of modern science, back in the sixteenth century, Latin was the language of scholarship. Any educated man could speak Latin; many could read Greek.

So making up words out of those languages was as though we were making up words out of English. The practice continued even as Latin and Greek slowly lost their place in the school curriculum — but for once inertia served a useful purpose. By keeping to the "dead" languages, the scientific vocabulary has remained very largely international. A large proportion of the various scientific terms are the same in Russian, for instance, as in English.

This is important since artificial barriers among the world's scientists will certainly slow scientific advance, and if each nation had its own nationalistic scientific vocabulary, there would be a scientific barrier at every language barrier.

If all educated people still had their Latin and Greek, each would see at once that *thermometer*, for instance, is a word that combines the Greek "therme" (heat) and "metron" (a measure). It is a "heat measure," and what could be plainer?

But there's a bright side to the loss. If, through the changing fashions in education, scientific words have become mysterious to us, there is the excitement of discovery in them. The *telephone* is just a telephone to us; we're so used to the word that we never give it a thought. But in Greek, "tele" means "far off" and "phone" means "voice." When we pick up the telephone, we pick up a "far-off voice." How can we describe the instrument more dramatically than we do whenever we casually name it?

In short, the scientific vocabulary is really an adventure. Hidden in the queer jawbreakers and in the shorter oddities are little stories; concise descriptions; thumbnail sections of history; tiny bits of testimony to great scientific achievements and to human error, too; reminders of great men and of mistaken and forgotten theories. As the words pass in review, they are all so different, and each in its own way is so interesting.

Far from frightening people away from science, the scientific vocabulary, looked at squarely and with understanding, should be one of the most powerful attractions of science. This book, I most earnestly hope, is evidence in favor of that view.

Absolute Zero

IN SOME cases, a "zero" from which to start counting can be set anywhere. Zero longitude (see MERIDIAN) is set arbitrarily. The same is true of zero in temperature. In the Celsius scale of temperature (see CENTIGRADE), zero degrees is set at the point where ice melts; while in the Fahrenheit scale, it is set at a point some way below the melting point of ice. In either case, though, temperatures below zero may be experienced.

Toward the end of the 1700's it began to seem that there might be a limit to cold. The French physicist Jacques A. C. Charles discovered in about 1787 that gases contracted 1/273 of their volume at 0°C. for each Centigrade degree they were cooled. (This is called *Charles's Law.*) If this were to continue, then, at about −273°C. the gas should disappear entirely. Of course, this does not happen. The gas invariably turns first liquid, then solid, as it is cooled.

The British physicist William Thomson extended the idea in the 1860's. He treated temperature as an expression of the velocity of movement of molecules in a substance. The colder the substance, the slower the motion until at a certain temperature (−273.18°C.) there was no motion at all. There could not be less than no motion, hence no temperature less than that. The temperature −273.18° was a real zero.

Now the Latin "solvere" means "to loosen" or "to free" and the prefix "ab-" means "from." Therefore, something that is "absolute" is something that is "freed from" all restraint. We speak of an absolute monarch, for instance. Anything that is an extreme can then be "absolute" since it is "freed from" all qualifications or maybes. You can have absolute opinions or be an absolute fool (or both). And a zero like −273.18°C. that is really zero is *absolute zero.*

Thomson's temperature scale, starting at absolute zero, is called the absolute scale, or the *Kelvin scale* (since Thomson was made Baron Kelvin of Largs in 1892). A temperature on that scale can be abbreviated either A. or K.

Academy

THE ANCIENT Athenian hero Theseus once carried off Helen of Sparta (who was later carried off by Paris, thus starting the Trojan War), and Helen's brothers, Castor and Polydeuces, went searching for her. Another Athenian, Academus, revealed her hiding place. For this reason, during the wars between Athens and Sparta, the Spartans (who held Castor and Polydeuces in special honor) always spared the site of the estate of Academus (about a mile northwest of Athens) whenever they invaded Athenian territory. This estate, the "Academeia," became a symbol of peace in a war-torn time.

Plato lived near the Academeia and resorted to that pleasant place with his students. There he taught for 50 years and his successors for another 800. The *Academy* (as we call it) was the most famous school of antiquity, and some schools still call themselves academies today as a result. Usually, the term is now applied to college preparatory schools.

The term *academic* is applied to anything pertaining to these academies, particularly to the type of learning they encouraged. Because Plato's philosophy was highly theoretical and abstract and did not concern itself with practical everyday matters, an "academic" question has now come to mean one that has no practical meaning, that is of theoretical interest only.

Other ancient philosophers also left the names of their places of teaching in the language. Aristotle used to teach in a gymnasium called the Lykeion, now called Lyceum. It was named in honor of a nearby temple to Apollo in his capacity as "wolf-killer." As such, he was called "Lykeios" probably from the Greek "lykos" (wolf). So *lyceum* is still used today for a school or a lecture hall. It is less popular than *academy* in America, but in France the word for what we call a high school is *lycée*.

Again, the philosopher, Zeno, taught in Athens at a place called Stoa Poikile ("painted porch" in Greek). His school of philosophy was termed *stoicism* because of it. Since Zeno taught that the way to happiness was to avoid undue emotion, a person who does not show his emotions is called a *stoic* to this day.

Acid

Sourness is one of the four fundamental tastes (the others being sweetness, saltness and bitterness) and it occurs naturally in such things as unripe fruits and some ripe ones. It was there that primitive man first became acquainted with the taste. In addition, it was already known in prehistoric times that certain liquids, such as milk, would develop sourness if allowed to stand. Fruit juices, on standing, fermented and became wine which, on further standing, also soured. The expression "sour wine" in Old French is "vin egre," and from this comes our own word *vinegar*.

The Latin word for "to be sour" is "acere." (The Old French "egre" is a form of that.) From "acere," two other words are derived: "acidus" as the adjective "sour," and "acetum" meaning "vinegar."

The medieval chemists were particularly interested in the sour substances. Strong vinegar could attack or corrode a number of metals and in its presence certain chemical changes took place that did not take place otherwise. About 1300, new and stronger chemicals of this sort were discovered. By their greater activity, metals and other substances could be brought into solution more quickly than by use of even the strongest vinegar. Virtually a chemical revolution took place.

All such compounds were named *acids* after their most prominent characteristic, sourness. Vinegar and fruit juices contained *organic acids* (see ORGANISM) while the new stronger substances which were obtained from nonliving sources were the *mineral acids*.

The particular acid in vinegar ("acetum," remember) was named *acetic acid* so that, as it happened, the two words of that name both came from the same Latin word "acere" and the substance bears a double dose of testimony to the fact that it is sour.

As in so many other cases, modern science has given the lie to the old name. To the modern chemist, an acid is any compound that has a tendency to lose a proton. If the tendency is great enough, the acid will be sour to the taste; if not great enough, it will not be sour, but it will still be an acid.

Adsorption

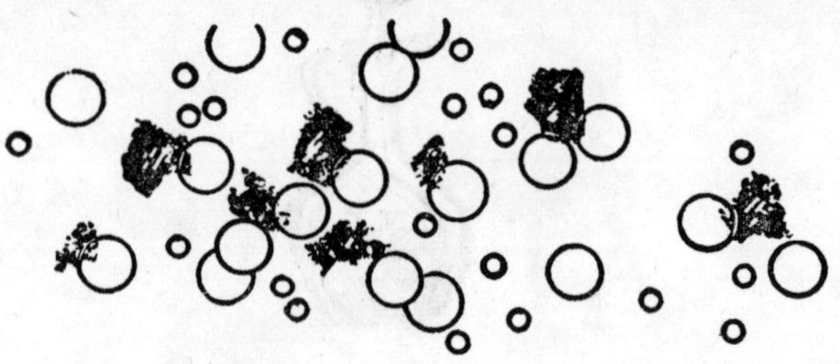

A SPONGE or a towel or a piece of blotting paper will take up water in defiance of gravity. The reason for it lies in the force of capillarity (see CAPILLARITY).

Before capillarity was understood, however, it looked as though the sponge, etc., were simply sucking up the water. (Sucking was one well-known way of lifting a fluid against gravity.) The phenomenon was therefore called *absorption*, from the Latin "ab-" (from) and "sorbere" (to suck up). A dry sponge, placed in a water-filled pan, "sucked water up from" the pan.

Chemists came across a similar phenomenon involving finely powdered substances. If a mixture of gases, for instance, were forced through a layer of finely powdered charcoal, some of the gas molecules would stick firmly to the surface of the charcoal particles and remain trapped there. The larger molecules would stick more tightly and the smaller ones would be shouldered aside so that they would get through the charcoal without sticking at all.

Poison gas vapors have larger molecules, generally, than do the oxygen and nitrogen of the atmosphere. Air containing poison gas, if forced through a canister containing powdered charcoal, will therefore lose its poison. The poison gas molecules will remain attached to the charcoal particles while the ordinary breathable air will get through. Such a charcoal canister, fitted into an airtight mask, is called a gas mask.

The poison gas has been "sucked out" of the air by charcoal as water is "sucked up" by a sponge. However, the reason is different. In the case of charcoal it is not capillarity that does the work, but all the many surfaces of the tiny particles to which the gas molecules adhere. Taking the "ad-" prefix (meaning "to") from "adhere," you have, instead of *absorption*, *adsorption*.

However, adsorption plays a part in capillarity, too, and sometimes it is difficult to tell whether a phenomenon should be truly spoken of as absorption or adsorption. Some scientists have suggested the word *sorption* to cover both types of phenomena.

Alcohol

Women, apparently, have for many centuries been darkening their eyelashes to make their eyes seem large and lustrous. Arabic women used a very finely divided powder for the purpose and the Arabic expression for that powder was "al koh'l" meaning "the finely divided."

The medieval chemist took up the expression, converted it to *alcohol* and used it for any fine powder, particularly a powder so fine it could not be felt; one that was *impalpable*, in other words, from the Latin "in-" (not) and "palpare" (to feel).

Some time in the early 1500's, chemists began applying the term to vapors that could be forced out of certain liquids. These vapors, you see, were impalpable, too. Wine, when heated, gave off a vapor which was first called "alcohol of wine" and then simply "alcohol."

When wine is heated, the alcohol it contains is more easily boiled than the water. The vapor is richer in alcohol than the wine originally was. If the vapor is cooled the resulting liquid is a stronger drink than the original. This process is called *distillation*, from the Latin "de-" (down) and "stilla" (a small drop). And, indeed, the vapors, when cooled, are collected as small cold drops falling into a container. Alcoholic drinks strengthened in this way are *distilled liquors* and the device in which the process is carried out is popularly called a *still*. Actually, distillation is an important chemical process that is used to separate individual compounds from many types of liquid mixtures.

The "-ol" suffix of the word *alcohol* is now used by chemists to name any compound which, like alcohol, contains a *hydroxyl group* in its molecule (so called because it consists of a hydrogen and oxygen atom in combination). The alcohol of wine contains also a two-carbon group in its molecule similar to that in ether (see ETHER), so it is called *ethyl alcohol* or *ethanol*.

Oddly enough, the Arabs call alcohol *spir't* from the English word *spirits*, so if we use their word for the purpose, they use ours.

Algol

Before the days of modern astronomy, any change in the changeless, perfect heavens was cause for concern (see comet). Even stars that merely varied in brightness from day to day could be regarded with alarm.

Actually there are many such variable stars, but only a few are bright enough and variable enough to be noted with the naked eye, and for some reason the ancients never commented on those few.

The most famous example is a star in the constellation Perseus, called *Beta Persei,* because at its brightest it is the second brightest star in the constellation (beta is the second letter of the Greek alphabet, so the star's astronomical name means "the second in Perseus." Other stars are named in similar fashion).

The variation isn't really much and it wasn't till 1669 that its light fluctuations were noted by a European. However, the star already had a significant Arabic name, *Algol.* (During the night of the western Dark Ages, the Arabs carried the main weight of progress in astronomy.) Algol comes from the Arabic "al" (the) and "ghul" (demon), so that the star is often called the Demon Star today, a sharp reminder of the fears its behavior gave rise to. As a matter of fact, "ghul" has come down to us as *ghoul.* Algol is "the ghoul."

But Algol is not really variable. It is an eclipsing binary (two stars revolving with an orbit edge-on to us). Every few days the dimmer star eclipses the brighter and the light reaching us is cut down.

A more remarkable star, in the constellation Cetus, is *Omicron Ceti.* (Omicron is the fifteenth letter of the usual Greek alphabet.) It actually pulsates and varies in brightness over irregular periods sometimes almost two years long. It can be as bright as the polestar or far too dim to be seen. The German astronomer David Fabricius noted it finally in 1596. By then astronomers were growing more sophisticated and were less troubled by odd events in the heavens. So Fabricius named it *Mira,* from the Latin "mirus" (wonderful). Though Algol is the "Demon Star," Mira, which is much more extreme, is the "Wonderful One." There seems no justice.

Almanac

ACTUALLY, MAN is a very long-lived creature. The only other creatures that may live a hundred years or more are certain trees and large turtles. At the other end of the scale, at least among animals visible to the naked eye, are certain insects which live their entire adult life in a day or less. (In their immature form, to be sure, such insects may live a total of one to three years.)

The common name for these short-lived insects is May fly, but the scientific name for the group to which they belong is *Ephemeridae*, from the Greek "epi-" (over) and "hemera" (day). Their lives are, indeed, over in a day.

But an *ephemeris* is something astronomical as well. It is a table or series of tables which gives the exact position of various heavenly bodies at certain times. This is valuable in navigation since position at sea can be plotted by observing the position of heavenly bodies. Since the table giving the position for a certain time is no longer useful once the time is past, it is good for the day only, so to speak, and hence its name.

Tables of astronomical data are also included in *almanacs*, a word sometimes used as a synonym for "ephemeris." The word almanac comes from the Arabic "al manakh" meaning "the calendar" or "the weather." The calendar and the weather are much the same, of course, since the seasons of the year come and go with the months. People have always believed in a more intimate connection, too, and have tied the weather to the phases of the moon, for instance. *The Old Farmer's Almanac* still supplies its readers with weather predictions for the entire year.

The very ephemeral nature of the almanac — that is, the fact that its tables grow out of date quickly — means that a new almanac must be put out periodically, usually every year. It is natural, then, to include information concerning the past year, such as news events and late statistics. As a result, our best-known almanacs these days have grown into a kind of annual one-volume encyclopedia.

Alpha Rays

THE DISCOVERY of radioactivity, which revolutionized science, came about this way: The French physicist Henri Antoine Becquerel was studying the way in which uranium salts glowed or "fluoresced" when exposed to sunlight. He wondered if the fluorescence contained X rays, which had just been discovered the year before (see X RAY), so he exposed the uranium salt to the sunlight and placed a carefully wrapped photographic plate nearby. Sure enough, the plate was fogged despite the wrappings.

However, on sheer impulse, he developed some wrapped plates which had been near some of the uranium compound in a dark drawer, and that plate was also fogged. It seemed that invisible radiation was being emitted by the uranium whether sunlight was present or not. What's more, this radiation, like the X rays, was much more penetrating than ordinary light.

In 1899, Becquerel (and others) noticed that some of the uranium radiation could be bent to one side by a magnet, so there were at least two different kinds of rays. Since their nature was unknown, it would be easiest just to call them rays A and B. The New Zealand-born British physicist Ernest Rutherford did just that, using the first two letters of the Greek alphabet, alpha and beta. He named them *alpha rays* and *beta rays*.

In 1900, the French physicist P. Villard discovered a new and particularly penetrating radiation emanating from uranium. These were automatically named *gamma rays* (probably by Rutherford, again) since gamma is the third letter of the Greek alphabet.

In that same year, Pierre and Marie Curie showed that beta rays consist of a stream of electrons moving at terrific speed. By 1909, Rutherford proved alpha rays to consist of streams of comparatively heavy particles, each consisting of two neutrons and two protons. As a result, speeding electrons are now called *beta particles*, while speeding two-neutron-two-proton combinations are *alpha particles*. (Gamma rays are not made up of particles, but are similar in nature to X rays, being, however, more energetic and therefore more penetrating.)

Amalgam

In the ordinary sense, an *alloy* is a foreign substance added to something that is otherwise desirable and added in such a way that it is not easily detectable. For instance, lead might be added to silver, water to milk, or sand to sugar. The word comes through the French from the Latin "ad-" (to) and "ligare" (bind). The impurity is bound to the original.

But an impurity can improve the original and metalworkers, particularly, realized this in very early days. A little bit of zinc added to copper produced a material called brass which was yellower than copper and more decorative. If tin were added to copper the result was bronze, a much harder material than either metal separately. In fact, before the introduction of iron, bronze was the hardest metal known. Armor was made of it (as is well described in Homer's *Iliad*) and the period was known as the Bronze Age. Even today, pure metals are hardly ever used. Instead metals are deliberately mixed to produce hundreds of new substances with desirable qualities not otherwise available.

So alloy has come to mean "a mixture of metals." So important is iron in our day that all alloys are divided into two groups, *ferrous alloys* which contain iron, and *nonferrous alloys* which do not. The stem "ferr-," used by chemists for naming compounds containing iron, comes from the Latin "ferrum" (iron), and the chemical symbol for iron is, as a matter of fact, Fe, for that reason.

The one other type of alloy that has a special group name are those involving mercury. Mercury is a liquid metal and its alloys are either liquids or soft solids. A soft solid metal is an odd phenomenon in a world which values metals for their hardness and toughness and this softness gave mercury alloys the name of *amalgams*. This is a corruption of the Greek word "malagma" meaning any soft, dough-like material. It is a silver amalgam (mercury plus several metals, mainly silver) and not silver itself that is used for "silver fillings" in teeth. The amalgam is soft enough to knead into the cavity, while chemical reactions harden it quickly on standing.

Amethyst

VARIOUS naturally occurring, but rare, stones are admired for a number of excellent reasons. They are beautiful to look at, and durable so that they don't become less lovely with time. The pleasure men get from such *jewels* is found in the very word itself which comes from the Old French "jouel" meaning "a little joy."

The romantic ancients, however, could not resist adding to the real virtues of jewels with occasional tales of magical properties. For instance one purple jewel was considered to be a remedy against drunkenness (perhaps because its color was like that of wine). Wine drunk out of a cup made of this jewel could never, it was said, intoxicate anyone. The Greek word for "intoxicated" is "methyein" while the prefix "a-" is used as a negative. The jewel that insures "no intoxication" is therefore an *amethyst.*

The Greek word "methyein" (intoxicated) in turn comes from "methy" (wine). Ordinary wine contains ethyl alcohol (see ALCOHOL) but a compound similar to ethyl alcohol can be obtained by heating wood in the absence of air. This second compound is quite poisonous and contains only one carbon atom in its molecule whereas ethyl alcohol contains two.

This compound from wood is called *wood alcohol,* for obvious reasons, or the classical equivalent, *methyl alcohol;* the word *methyl* coming from "methy" (wine) and "hyle" which means "matter" generally, but can be used to mean "wood" in particular. *Methyl* is the "wine from wood."

Chemists use the "meth" stem for other atom groupings containing a single carbon atom. A gas known as marsh gas (it is found over marshes, where it is formed from decaying vegetable matter) possesses a molecule made up of one carbon atom and four hydrogen atoms. Its proper chemical name is therefore *methane,* the "-ane" suffix being reserved for certain types of *hydrocarbons* (compounds with molecules made up of hydrogen and carbon atoms only).

Ammonia

One of the more important of the gods of ancient Egypt was the deity Amen or Amun, who was the patron god of the Egyptian city of Thebes on the upper Nile. When Greek culture spread throughout the Near and Middle East after the conquests of Alexander the Great, the Greeks took to identifying their own gods with those of the other people with whom they mixed. For instance, they had already identified their own chief god, Zeus, with Amen (or Ammon as they spelled it) and a temple to Zeus-Ammon was built in an oasis in the North African desert.

Any desert area has a problem when it comes to finding fuel. One available fuel in North Africa is camel dung. The soot that settled out of burning camel dung on the walls and ceiling of the temple contained white saltlike crystals, which were then called *sal ammoniac*. The word "sal" is Latin for "salt" so that the phrase means "salt of Ammon."

At various times, in the centuries following, a pungent gas was obtained from sal ammoniac but Priestley (see OXYGEN), in 1774, was the first to collect the gas separately and study it. He called it *alkaline air* because it would dissolve in water and then exhibit alkaline properties (see POTASSIUM). However, the name *ammonia*, from "sal ammoniac" won out and is the name of the gas to this day.

The ammonia molecule is made up of three hydrogen atoms and one nitrogen atom. If a fourth hydrogen is added, the *ammonium ion* is formed and it is this that forms saltlike compounds. (Sal ammoniac turned out to be *ammonium chloride.*)

The ammonia molecule minus a hydrogen atom is an *amine group*. If instead of the missing hydrogen, a carbon-containing group of atoms is attached, the resulting compound is an *amine*. Proteins are made up of long strings of relatively simple compounds containing both amine groups and acid groups. These compounds are therefore called *amino acids*, so it turns out that in the most important substances in our bodies, we carry a reference to the great god Amen of Egypt.

Amphibious

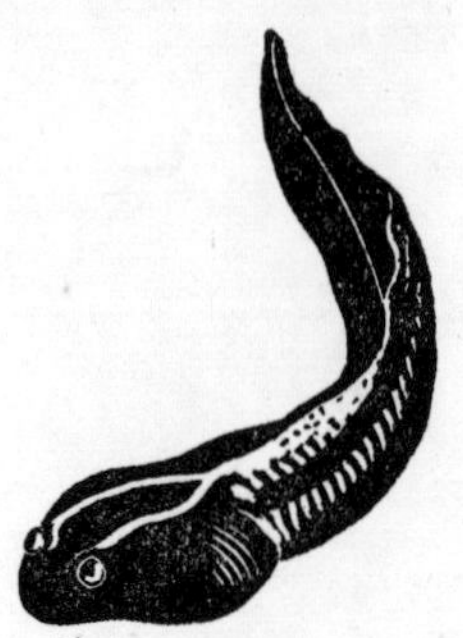

THE FIRST backboned animals to emerge from water and take to breathing air and living on land were ancestors of today's frogs and toads. (They were preceded by insects and snails, probably, but those were animals without backbones.)

These frog-ancestors did not make a complete change-over but emerged on land only in adult life. Their young had to pass through an immature stage in water, living a fishlike existence, before they, too, exchanged gills for lungs and became air-breathing. Because these creatures lived part of their lives in water and part on land, they are *amphibious* from the Greek "amphi" (on both sides of) and "bios" (life). They led a double life.

(During World War II, this word was also used for attacks made by land and sea, which is reasonable enough. However, when attacks by air, land and sea were made, the corresponding word should have been "tribious" (three lives). However, newspapers insisted on calling such a triple attack "triphibious," a horrible word to anyone who knows his derivations.)

The most common amphibians today are frogs and toads, both of which words come from the Anglo-Saxon. The immature stage of either creature is a *tadpole*. The prefix "tad-" is just a corruption of "toad." The "pole" is a corruption of "poll," which is an old-fashioned word for "head." (This word still lingers as a synonym for taking a vote, since that amounts, so to speak, to counting heads on both sides of a question. We also have *poll tax* which is a tax laid on the head of each household.) In any case, a tadpole seems to be little more than a head swimming about, and that is what it means: "toad-head."

A tadpole is called, even more dramatically, a *polliwog* also. Again "poll" means "head" while the "wog" portion of the word is a corruption of "wiggle." A polliwog therefore is a "wiggling head" and next time you see one, see if that isn't an exact description.

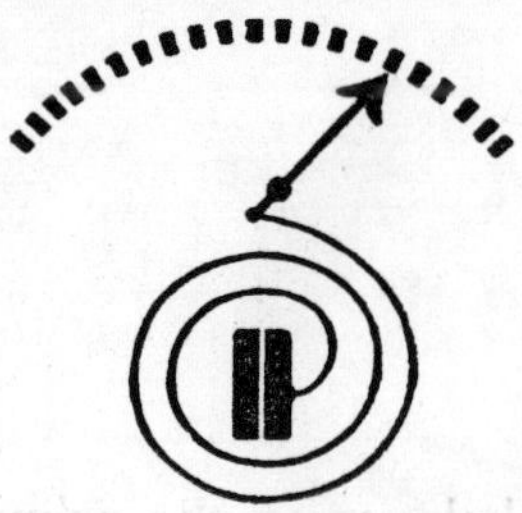

Aneroid

THE MOST FAMILIAR type of barometer involves a long glass tube full of mercury (see BAROMETER). While this may be very useful, it is not easily portable.

Another way of measuring air pressure is to make use of a thin-topped, disc-shaped metal box, hollow and evacuated. The outer air presses against the thin top and forces it in somewhat. The greater the pressure, the more the top is pushed in, and vice versa.

Naturally, the amount of movement of the thin sheet of metal is not much, but it is connected within by a system of levers that magnifies the motion and passes it on to a coiled spring that in turn swings a pointer on the outside of the box. The pointer marks the air pressure as so many inches (or millimeters) of mercury, according to the distance the air pressure has made it swing.

Since such a barometer does not involve the use of a liquid, it is called an *aneroid* barometer, from the Greek "a-" (general negative) and "neros" (wet). It is a "not wet" barometer.

Since air pressure decreases with increasing height according to a known rule, an aneroid barometer will tell you how high above sea level you are (if allowance is made for weather conditions, since air pressure is lower on stormy days than on fine days). In that case, the instrument becomes an *altimeter*, from the Latin "altus" (high) and "metrum" (measure).

The suffix "-meter" can be used with either Latin or Greek prefixes since "metrum" is Latin and "metron" Greek, both meaning "measure." Many names for scientific instruments include it. Perhaps the most familiar example is the thermometer, which measures temperature (see CENTIGRADE). More glamorous examples are the instruments put into artificial satellites which make various measurements in the fringes of the atmosphere and make the results known to us by appropriately varying radio signals. These are examples of *telemeters* (from the Greek "tele," meaning "distant"). They are devices that "measure at a distance."

Angiosperm

Most of the plant life in the world belongs to the primitive Thallophyta (see PLANKTON), which are mainly sea plants. The most advanced plants (and those most familiar to us) are the land-dwelling *spermatophytes*. These have roots, leaves, flowers, fruit, seeds — all the things we usually think of in connection with plants. In fact, the Greek "sperma" means "seed" while "phyton" means "plant," so that the spermatophytes are "seed-plants," as opposed to thallophytes and certain of the more primitive land plants such as mosses and ferns which do not produce seeds.

The spermatophytes include, first, the *gymnosperms*. In these plants, the *ovule* (that is, the object which will, after pollination, become the seed — the word being derived from a Latin form, "ovulum," meaning "little egg") is exposed on the surface of the organ that forms it. The Greek "gymnos" means "naked" so a gymnosperm is a "naked-seed" plant. The gymnosperms include the various evergreens, with their needle-shaped leaves that are not shed in the fall and their woody fruit, called *cones* because they are often conical in shape.

The remaining spermatophytes are the *angiosperms* from the Greek "angion," meaning "vessel," because the ovules when formed are enclosed in a vessel called the *ovary* (from the Latin "ovum," meaning "egg") into which the pollen must penetrate to achieve pollination. All the common flowering plants and deciduous trees belong in this group. A *deciduous* tree is one that sheds its leaves in the fall, from the Latin "de-" (down) and "cadere" (to fall); the leaves "fall down."

When the seed develops, the first tiny leaves to appear lie in a small hollow in the seed. They are called *cotyledons*, a Greek word meaning "a cup-shaped hollow" (from "kotyle," a kind of cup used by the Greeks). The angiosperms are divided into two groups, depending on whether they possess one cotyledon or two. The *monocotyledons* (the Greek "monos" means "alone" or "solitary") include all the grains and grasses as well as some flowers such as lilies and orchids. Most angiosperms, however, have two such leaves and are *dicotyledons* (the Greek "di" means "twice").

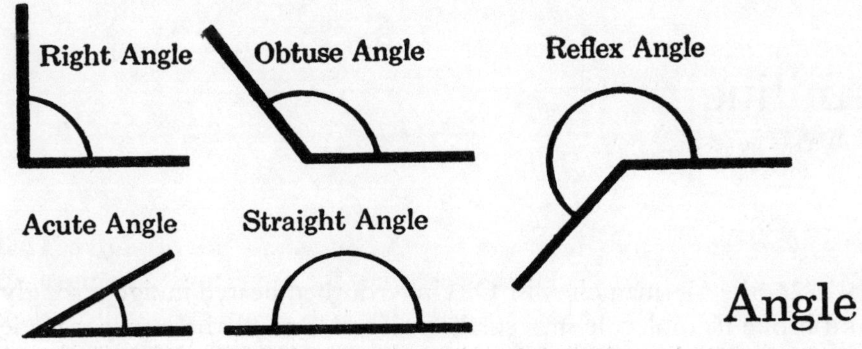

Angle

GEOMETRY (Greek "ge," earth; "metron," measure) concerns itself largely with the intersection of straight lines. Two lines intersecting form an *angle*, a word taken from the Latin "angulus" meaning a "corner." The most familiar angle is the *right angle* (see PERPENDICULAR) and the ordinary street corner is a right angle indeed, in a city laid out like Philadelphia or New York.

When two lines meet in such a way as to form a sharper point than a right angle does, it is called an *acute angle*. In Latin, "acuere" means "to sharpen," and the past participle, "sharpened," is "acutus." If the lines meet in such a way as to form a blunter point than a right angle does, it is called an *obtuse angle*. The word *obtuse* is taken from the Latin prefix "ob-" meaning "against" and "tundere" meaning "to strike." The past participle of the verb "obtundere" is "obtusus." If you strike against a sharp edge, you see, you blunt or dull it; you make it obtuse.

If the two lines that form the angle become so obtuse that they meet in the same straight line, there is no angle in the ordinary sense, but the geometers, to be complete, find it convenient to call a straight line a *straight angle* sometimes. (The word *straight* comes from the Anglo-Saxon word "streccan" which also gives us our word *stretch*. Stretch a cord, after all, and you make a straight line out of it.)

An angle can be still larger than a straight angle; one arm bends beneath the horizontal and is on its way back to the other arm, so to speak. This is a *reflex angle*, from the Latin "re-" meaning "back" and "flectere" meaning "to bend." "Bent back" in Latin is, in fact, "reflexus."

Aniline

In 1826, the German chemist O. Unverdorben heated indigo strongly and broke its molecule into smaller pieces. One of these pieces made up a new nitrogen-containing organic liquid. In 1840, the method was improved and the name *aniline* proposed for the new compound from "anil," a name of the indigo plant (see INDIGO). Aniline could also be obtained from coal tar (a pitchy substance derived by heating soft coal in the absence of air).

In 1856, an eighteen-year-old British chemistry student, William Henry Perkin, was trying to make quinine (an antimalarial medicine), out of simpler chemicals. The makeup of the quinine molecule was not known at the time so the chances of success were very minute and, of course, Perkin failed.

But what a failure! In the course of his experiments, Perkin treated aniline with various chemicals (because he thought, mistakenly, that the aniline molecule resembled the quinine molecule) and got a black, gooey mess for his pains. Undoubtedly, he ought to have dumped it but there was a purple glint to it and he began thinking. He sent some of it to a dye establishment and they were interested.

Perkin dropped everything and concentrated on getting the purple dye out of the black mess and learning how to make more of it. He discovered a more economical way of making aniline from coal-tar chemicals and set up a factory. The dye was named *Aniline Purple*, but the French dyers who took up the new material invented the word, mauve, for the color, because it resembled in color the flower of the mallow (Malvus sylvestris). The dye was also known as *mauveine*.

Mauveine was the first of hundreds of synthetic dyes produced by the chemical industry. As a class, these are called *aniline dyes* or *coal-tar dyes*. They have put the natural dyes out of business, including indigo, from which aniline was first obtained and after which it was named. And Perkin lived an additional half century, rich and famous, because he had the wit not to tip a mess down the sink and the drive to construct an industry out of it.

Anthracite

Bitumen is the Latin word for a kind of tarry material that was useful in that it was soft enough to be smeared on objects, sticky enough to remain and harden there, and capable of making the objects so treated waterproof. Bitumen is referred to in the Latin version of the Bible and is translated as "pitch" or "slime" in English. Noah's ark was coated with bitumen and so was the ark of bulrushes that saved the infant Moses.

Now coal is mostly carbon, having been formed from plant life that grew in eons past. During the course of ages, most of the hydrogen, oxygen and nitrogen atoms in the plant tissues were lost, leaving the carbon. Most coal, if heated in the absence of air, will give up what is left of these other atoms in various molecular combinations as gases and vapors. Some of these vapors can be cooled into a black bitumen-like material called coal tar, so the kind of coal which can be so treated is called *bituminous coal.* It yields bitumen, in other words.

A less common variety of coal (found particularly in eastern Pennsylvania) contains so much carbon (90 per cent or more) and so little of anything else, that it produces no bitumen to speak of. It burns with more heat and less smoke than does bituminous coal and is therefore the variety that is used in home-heating. It is *anthracite coal*, from the Greek "anthrax" which means "coal," so that, in a sense, anthracite coal is "coal coal."

One of the compounds obtained from coal tar has a molecule made up of 14 carbon atoms in a triple ring. It is called *anthracene*, again from "anthrax." *Anthrax* itself is the name given to a fatal disease attacking domestic animals and sometimes man. One of the symptoms in man are the coal-black infected pustules that appear, hence the name.

On similar principles, the word *carbuncle* (from the Latin "carbunculus," meaning "a little coal") can be applied to a red garnet or to a large boil, because both resemble a smoldering coal. I imagine that the boil makes the more convincing imitation to a person unfortunate enough to have one.

Antibody

Citizens of ancient Rome owed their governments certain "gifts." These might be of money (as we pay taxes) or of services (as we go off to war). Certain citizens would, for one reason or another, be exempt (just as today some organizations are tax-exempt and certain people are draft-exempt).

The Latin word for these services or obligations is "munia" and the Latin prefix "im-" means "not." A person who was not expected to make a particular gift of money or services was said in Latin to be "immunis." In English, the word has come down as *immunity*.

We all know that a person who has had measles, or certain other diseases, will practically never get it a second time. He has become *immune* to that disease.

The body achieves immunity in its attempts to battle the first attack of a disease. The body then begins to form protein molecules in the blood that are especially designed to combine with the disease germ and render it harmless. Or the molecules might combine with a poisonous compound produced by a germ and neutralize its action. (Such germ-produced poisons are called *toxins* from the Greek "toxon," meaning "bow," because arrows were often poisoned with what the Greeks called "toxicon pharmakon" or "poison of the bow.")

The germ-fighting proteins linger in the blood after the person recovers from a disease such as measles, and a second attempt at infection is met instantly by these ready defenders. The defense proteins are called *antibodies* from the Greek "anti" (against) since they are bodies (i.e., substances) formed to act against germs. The germs or their poisons (or anything in fact that causes the production of antibodies in the first place) are *antigens*, the Greek suffix "-gen" meaning "to produce."

Appendix

Living creatures, including man, are virtual museums of structures that have no useful function but which represent the remains of organs that once had some use. (Man has tiny bones once meant for a tail, and unworkable muscles once meant to move the ears.)

The small intestine, for instance, enters the large intestine about two or three inches from its lower end. That lower end is therefore a kind of blind alley called the *caecum*, from the Latin "caecus" (blind). At the end of the caecum is a much narrower tube about three to four inches long. In general, structures that hang from a part of the body are called *appendices* (singular, *appendix*) or *appendages*, from the Latin "ad-" (to) and "pendere" (to hang). They are structures that "hang on to" the body.

This narrow tube (also a blind alley) is just long and narrow enough to resemble a worm, so that it is called the *vermiform appendix*, from the Latin "vermis" (worm) and "forma" (form) — the "appendix in the form of a worm." However, such is its notoriety that this particular organ is usually called simply the *appendix*.

In certain plant-eating animals, the caecum is a large storage place where food may remain to be broken down by bacteria so that the animal itself may more easily digest and absorb it. The appendix in man and the apes (it occurs in almost no other animal) is what remains of that large caecum. It indicates that the fairly near ancestors of man and the apes were plant-eaters. The appendix is thus the useless remainder of a once useful organ; it is a *vestige*, from the Latin "vestigium" (footprint). Just as a footprint is a sign that a man once passed that way, so a vestige is a sign that a useful organ once passed that way.

Every once in a while the appendix is worse than useless. It becomes inflamed and must be cut out lest it agonize and kill its owner. The condition is *appendicitis;* the operation, an *appendectomy*. The suffix "-itis" is from the Greek and is used to mean "an inflammation of," while the suffix "-ectomy" comes from the Greek "ektome" meaning "cut out."

Aqua Regia

THE ALCHEMISTS of the Middle Ages used colorful language to describe the materials they worked with. Liquids, for instance, were generally named as some descriptive type of "aqua" (which is Latin for "water").

Thus, when it was first learned how to distill wine and get alcohol, a watery solution containing enough alcohol to burn was obtained and this was called *aqua ardens* which is Latin for "burning water." A solution containing larger quantities of alcohol was called *aqua vitae* ("water of life"), probably because of the new life that drinking it appeared to give people who were feeling a bit on the dragged-out side. The expression is still used for various forms of brandy and similar liquors.

Sometime in the thirteenth century, the alchemists discovered the strong mineral acids. This was a milestone in chemical history since acids can be used to dissolve many things that will not dissolve in water. The strongest known acid in ancient times was vinegar (see ACID) but the mineral acids were millions of times stronger and made many chemical reactions and processes possible that were not possible before.

Nitric acid, for instance (see RAYON), was discovered and named *aqua fortis* ("strong water") and so it was, for it ate away almost any substance with which it came in contact, including all the metals then known except gold. If hydrochloric acid (which was discovered three centuries later) or ammonium chloride were added to the nitric acid, the mixture turned green and the acid became still stronger, for now it actually dissolved gold. (This was because the hydrochloric acid reacted with the nitric acid to form the green element, chlorine, which was what attacked the gold.)

Since gold was the king of metals, an acid that would dissolve it must surely be the king of waters, and so it was named *aqua regia* ("royal water"). Though almost all the names of alchemy have died out, this one persists and is used to this very day for a 1-to-4 mixture of nitric and hydrochloric acids. (It dissolves platinum, too.)

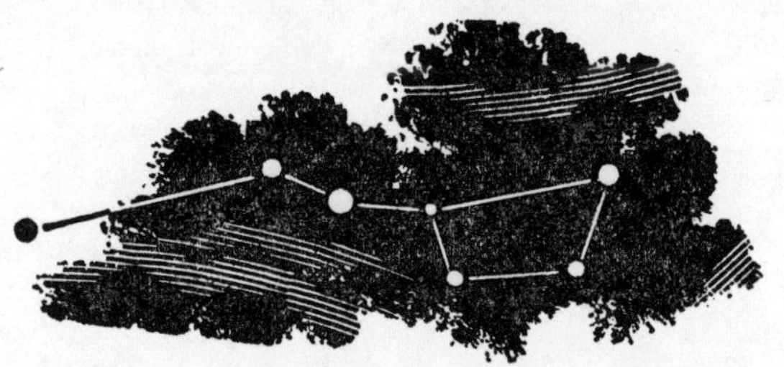

Arctic

THE EARTH'S AXIS of rotation is inclined 23½ degrees to its plane of revolution. On December 21, then, when the North Pole is inclined the full 23½ degrees away from the sun, all points within 23½ degrees of the pole go at least one full day without seeing the sun. At the same time, the South Pole is inclined the full 23½ degrees toward the sun so that all points within 23½ degrees of that pole will enjoy uninterrupted sun for at least one full day. On June 21, when the North Pole is inclined toward the sun and the South Pole away from it, the situation is reversed.

Now as you go farther north from the equator, the stars of the northern sky seem to climb higher in the sky. Eventually, the most prominent constellation of the northern sky, the "Big Dipper" or "Great Bear" would be overhead at some time of the night. The Greeks, therefore, referred to the north as the *arctic*, from "arktos" (bear). It was the region where the "bear" was overhead.

Naturally, the southern regions were the *antarctic*, from the Greek "anti-" meaning "against" or "opposite to." The south, after all, was in the direction opposite to the north.

In modern geography, the arctic refers to that portion of the earth's surface lying within the 23½-degree limit of the North Pole while the antarctic is the similar region about the South Pole. The imaginary circle that limits this region is the *Arctic Circle* in the north and the *Antarctic Circle* in the south. Furthermore, the ice-covered continent lying almost entirely within the Antarctic Circle is *Antarctica.*

The Great Bear itself is known to astronomers by the Latin name *Ursa Major* (from the Latin "ursus," meaning "bear," and "major," meaning "larger"), but the Greek word comes into play in connection with one of the stars, a bright one in the constellation Bootes. It is near Ursa Major and seems to be a gleaming eye keeping perpetual guard over the bear. It is named *Arcturus,* from the Greek "arktos" (bear) and "ouros" (guard).

Argon

THERE IS an important difference between iron and gold. Iron rusts and crumbles away. Gold does not. The reason is that iron will combine with the oxygen and water vapor in air to form *rust* (from an Anglo-Saxon word meaning "red"). Gold, on the other hand, does not combine with oxygen or with scarcely any substance except under extreme conditions. It is an inert element (see INERTIA).

To the ancients, however, there seemed another explanation. Standoffishness was characteristic of aristocrats. Noblemen, after all, didn't associate with just anyone, but only with their equals. The higher the nobleman, the fewer his equals and the more standoffish he was. Gold, obviously, was a "noble metal."

Now in 1894, the Scottish chemist William Ramsay discovered a gaseous element that made up 1 per cent of the air. It would not combine with other elements under any conditions. Its atoms would not even pair up with each other. It is one of the group of *inert gases*, for that reason (also called *rare gases*, because present in air in such small quantity).

Ramsay called this particular gas *argon*, from the Greek "a-" (no) and "ergon" (work). It was too lazy to do the work of combining with other substances. But the old notion of "nobility" persisted. Here was something even more standoffish than gold itself, and so argon was called a "noble gas." (Or perhaps that's not so odd. A hereditary nobility tends to become lazy at that.)

Ramsay proceeded to find four other inert gases in the air in the next four years. They were even rarer than argon. One was *helium*, found years before in the sun (see HELIUM). Then there was *neon* (from the Greek word for "new" which is "neos"), *krypton* (from the Greek "kryptos" meaning "hidden") and *xenon* (from the Greek "xenos" meaning "stranger"). After all, these gases were new and strange and had been hidden in air a long time before their discovery. (A sixth and last member of the series was discovered later and named *radon*. See RADIOACTIVITY.)

Artery

In ancient times, only one type of blood vessel was recognized. In Latin, the word for it was "vena" and in English it is *vein.*

There are other vessels like the veins but, in general, larger and with thicker walls. When anatomists examined these in dead bodies, they found them to be empty. They assumed that they carried air to various parts of the body and were therefore extensions of the windpipe.

The Greek name for the windpipe was "arteria," perhaps from "aer" (air) and "terein" (to keep); a vessel in which air was kept. This name was transferred to the empty veinlike vessels which are now called *arteries.* (The windpipe is no longer known by its old name but is called the *trachea,* from the Greek "trachys" (rough), because it has circular stiffenings of cartilage which make it rough to the touch, as you will find if you feel your throat.)

The Greek physician Galen, in the second century A.D., was the first to find that the arteries carried blood as did the veins, but for many centuries afterward, no one understood properly why two sets of vessels were needed. In 1628, the English physician William Harvey set forth the principle of the circulation of the blood; that it did not rest or oscillate back and forth in the vessels, but flowed in a single direction, leaving the heart via the arteries and returning via the veins.

There was even then no known connection between arteries and veins, yet this was necessary if circulation were to take place. In 1661, four years after Harvey's death, the Italian physiologist Marcello Malpighi supplied the missing link. He was the first to use the microscope systematically on plants and animals and, among other things, he saw (in frogs first) tiny vessels, thinner than hairs, through which blood flowed from arteries to veins, bathing all tissue cells as it did so. These tiny thinner-than-hair vessels are *capillaries* from the Latin "capillus," meaning "hair."

Ascorbic Acid

UNTIL FAIRLY RECENT times, scurvy (medieval Latin name "scorbutus," of uncertain origin) was a serious human affliction. It began with weakening and with muscular pains, then continued with sore gums and hemorrhages. Eventually teeth fell out, hemorrhage grew more severe, and death was the end. Looking back from the greater knowledge of today, it seems people ought to have noticed the connection of the disease with diet earlier than they did.

For instance, it struck whenever diet was monotonous and lacked fresh fruit and vegetables — on long sea voyages particularly; in hard-pressed armies; in besieged cities; in prisons and poorhouses.

Scurvy will not strike people under any of these conditions if certain fruit juices are available. After several decades of sporadic experimentation, the British navy, in 1795, began to force their sailors to drink a daily ration of lime juice. The sailors must have complained bitterly, but scurvy stopped. (And to this day, British sailors are referred to slangily as "limies.") The other citrus juices, tomato juice, and various fresh vegetables are also effective in preventing scurvy.

By 1907, biochemists began suggesting in print that there might be a chemical in these foods which was needed by the body and that scurvy was the result of the absence of this substance. About that time, chemicals called vitamins began to enter the thinking of nutritionists (see VITAMIN), so the hypothetical anti-scurvy, or *antiscorbutic,* chemical was eventually named *vitamin C.* (The letters A and B were already taken.)

The Hungarian biochemist Albert Szent-Györgyi isolated a chemical from cabbage in 1928 which the American biochemist Charles G. King showed in 1932 was the chemical needed by the body to prevent scurvy; it was named *ascorbic acid.* (The Greek prefix "a-" is a general negative, so the compound is the "no-scurvy acid.")

The American Medical Association frowns on chemical names that mention the disease against which they are effective since this may encourage self-dosing. They have suggested *cevitamic acid* as an alternative (from "C vitamin") but this name just hasn't caught on.

Asteroid

A PLANET IS an astronomical body which revolves about the sun, but there are other astronomical bodies which revolve about planets. The moon, for instance, revolves about the earth (and is carried by it about the sun, so that it revolves about the sun, too). Most of the other planets have smaller bodies circling them.

Such smaller bodies may be called *moons* in analogy to our own moon, but a more frequently used word is *satellite* from the Latin word "satelles," meaning "attendant" (since the moon attends its planet on the latter's journey about the sun). The Russian word is *sputnik*, meaning "one who travels with another."

Beginning in 1801, hundreds of tiny planets were discovered circling the sun in orbits lying between those of Mars and Jupiter. They circled the sun, so they were planets. They were so small, however (even the largest is only 480 miles in diameter, compared with a minimum of 3000 miles for the other planets), that some new name seemed called for.

The most frequently used name is *asteroid* from the Greek word "aster," meaning "star," and the Greek suffix "-oeides," meaning "having the form of." These small planets have the form of stars when seen through the telescope instead of showing visible discs as the other planets do. In actual fact, however, they are completely unlike stars, so many people prefer the alternate name *planetoid*.

Even planetoid isn't quite a fair name. The planetoids do not merely have the form of planets; they are planets. To emphasize their small size, though, they are frequently called *minor planets* and that is perhaps the best name of all.

Tiny minor planets that enter the atmosphere of the earth and burn up are called *meteors*, from the Greek word "meteoron," meaning "a heavenly phenomenon." If a meteor is not entirely consumed in the atmosphere, the part that survives and hits the earth is a *meteorite*. While it is still in space and before it enters the atmosphere it is a *meteoroid*. If it is of microscopic size, as billions upon billions are, it is a *micrometeor*.

Atmosphere

A SPHERE (from the Greek "sphaira," a ball) is a solid, the surface of which curves equally in all directions. The earth's shape is roughly that of a sphere but not quite. It is slightly flattened at the poles so that the curvature is less sharp there than at the equator. It is a *spheroid,* but under ordinary circumstances, it is still spoken of as a sphere.

The solid matter of the earth itself, for instance, is the *lithosphere,* from the Greek "lithos" (stone). Enclosing three fourths of the lithosphere is a shell of liquid making up the earth's ocean. If the lithosphere were to disappear, the oceans would be seen to form a sort of hollow (and incomplete) sphere. They are referred to, then, as the *hydrosphere,* from the Greek "hydor" (water). Outside both lithosphere and hydrosphere is a hollow sphere of gas, the *atmosphere,* from the Greek "atmos" (vapor).

The atmosphere grows rapidly thinner as it extends outward from the earth but very thin wisps reach outward for hundreds of miles and it has no definite end. As the atmosphere thins, various layers have different properties and have received different names.

About 75 per cent of the atmosphere lies within seven miles of the earth's surface. In this lowest layer are the clouds and storms and all the weather changes we witness. It is the *troposphere,* from the Greek "tropos" (change). The ten miles or so above the troposphere consists of a layer of air called the *stratosphere.* This is from the Latin "stratum" which is the past participle of "sternere" (to spread). The theory was that in the absence of storm and change, the air simply "spread" out quietly, perhaps in several sub-layers.

For several hundred miles above the stratosphere is the *ionosphere,* so called because layers of ionized gas (see ION) occur within it, the ions being formed through the action of the short-wave radiation of the sun. Finally, from a height of 300 miles or so, indefinitely upward, is the *exosphere,* from the Greek "exo" (out). It is the part of the atmosphere that is at the outer limit, you see.

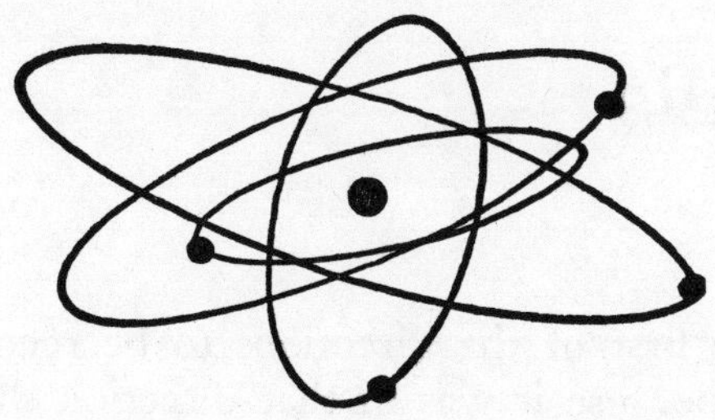

Atom

The ancient Greeks were much interested in speculating on the nature of the world about them and succeeded in evolving many fascinating theories as a result. They were usually wrong, if we judge them by what we now believe to be true — but not always.

For instance, two Greek thinkers, Leucippus of Miletus and Democritus of Abdera, decided that substances could not be broken up into smaller and smaller particles indefinitely. Eventually, they thought, particles would be obtained so small they could be divided no farther. There were a number of varieties of such particles, each making up a different substance. By combining them in different ways, still other substances would result. The Greek word for "indivisible" is "atomos," and so these particles were named after the fact that they could not be divided. They were called *atoms*.

This theory did not win favor among the Greeks, but over 2000 years later, it was resurrected. In 1803, the British chemist John Dalton decided that the facts uncovered by the still-new science of chemistry could best be explained by supposing every chemical element to be formed of tiny indivisible particles. Each element had its own characteristic type of particle and by varying the manner of combination of these, all existing substances could be constructed.

Dalton, following the old Greek theory, named his indivisible particles *atoms*, and this time the atomic theory met with approval.

Strangely enough, however, it was discovered in 1896 that atoms are not indivisible after all. Certain complicated atoms, it was found, broke up spontaneously, liberating particles far smaller than atoms. Then, scientists learned how to break up atoms in the laboratory. Now, man's whole future hinges upon the manner in which atoms break up and fuse together and on the behavior of particles smaller than atoms. But still the name is atom — "indivisible."

Aurora Borealis

East was probably the first of the directions to be recognized by primitive man, if only because it was in that direction that the sun rose. Huddled in the cold and danger of the long winter night, he must have waited eagerly for the first signs of the returning sun with its light and warmth; and he must have learned in which direction to watch for it.

The Sanskrit word "usas" was used for that direction, apparently derived from another word meaning "shining." And from that came the Greek "eos" and the Latin "aurora" (which may have originally been "ausosa"). The words mean both "dawn" and "the direction of the dawn." Our word *east* obviously comes from "eos," or at least the two have a common ancestor. (Similarly, *orient*, which also means "east," comes from the Latin "orior," meaning "to rise." Latin for "rising" is "oriens.")

There is one kind of night light or "dawn," however, which did not seem to have any connection with the sun. (Actually it does, but the ancients couldn't know that.) Every once in a while an upheaval on the sun's surface sends a stream of electrons shooting into space. When this stream strikes the upper regions of the earth's atmosphere, it pumps energy into the atoms and molecules of the air and causes them to glow. A beautiful many-colored light in moving sheets and streamers is formed. The light derived its name, however, from that which seemed most remarkable of all to the early observers — its direction.

The electrons, being electrically charged, are deflected by the earth's magnetic field and strike the atmosphere in the polar regions mainly. From the northern hemisphere, the light is seen in the north, and is called the *Aurora Borealis* (Latin for "northern dawn," since Boreas was the word in both Latin and Greek for the north wind). It is usually called the *Northern Lights* in English. How different from the much more usual "eastern dawn"!

In the southern hemisphere, the light is seen in the south and is called the *Aurora Australis* ("southern dawn," since Auster is the south wind in Latin).

Bacteriophage

VIRUSES, WHICH are organisms so small that they cannot be seen by the ordinary microscope (see VIRUS) are the cause of some of our most notorious diseases: colds, influenza, measles, mumps, smallpox, yellow fever and poliomyelitis. There are also the *plant viruses*, so called because they infect plants. One of these, the tobacco mosaic virus, was the first of all viruses to be experimented with, and the first to be isolated.

There are even viruses which infect bacteria and are parasitic upon them. In 1915, F. W. Twort first noticed that certain colonies of the bacteria he was growing became translucent and seemed to melt away. If he made an extract of the colonies and filtered it, the filtered extract, when added to normal colonies of bacteria, caused them to begin fading, too. The Canadian investigator Félix Hubert d'Hérelle extended such studies in 1918, suspected the existence of a virus, and called it *bacteriophage*. The suffix "-phage" comes from the Greek "phagein" (to eat), so that a bacteriophage is a "bacteria-eater" which is exactly right. Oddly enough, although they are parasites on simple one-celled creatures, the bacteriophages are larger and more complicated than most of the viruses that infest multicellular plants and animals.

There are creatures intermediate in size between bacteria and viruses. Like the viruses, they can grow only within the living cell and not on nonliving media as bacteria can. They are large enough, however, to be made out by ordinary microscope as small bodies within infected cells. These were called first *rickettsial bodies*, then simply *rickettsia*, after the American pathologist H. T. Ricketts who first discovered them in connection with the disease Rocky Mountain Spotted Fever. They are now known to cause a number of diseases (including typhus), all of which are spread by such creatures as ticks and lice. The use of DDT to kill these vermin just about defeats these diseases.

Ballistics

When an object is thrown, it follows a certain path under the influence of the force of the throwing arm and the force of gravity. The path it follows is its *trajectory*, from the Latin "trans-" (across) and "jacere" (to throw); it is the path resulting, in other words, when an object is thrown across from here to there. The thrown object itself is a *projectile*. The Latin prefix "pro-" means "before" or "forward" so that a projectile is something "thrown forward." It may also be called a *missile*, from the Latin "mittere" (past participle, "missus"), meaning "to send" or "to throw." (If it disturbs you, by the way, that "jacere" and "mittere" both mean "to throw," just think of the English words *throw*, *toss*, *hurl*, *cast*, and *pitch*.)

When projectiles are small and thrown weakly, their paths may be estimated in advance quite well, even though that path is curved and not straight, and the thrower can make proper allowance. Thus, a baseball pitcher may have excellent control and a boy may be fiendishly accurate with a slingshot.

However, once cannon were invented and heavy lumps of stone and iron were hurled over long distances, mental estimates were no longer sufficient. Trajectories had to be studied mathematically, and the science was named *ballistics* from the Greek "ballein" (to throw) — which is possibly related to the word *ball*.

In a perfect vacuum, a missile would follow a perfect parabola (see PARABOLA) as its trajectory and this would be easy to compute. However, air resistance slows it in its flight and changes its trajectory. Since air resistance varies with the height of the missile above the earth, with wind velocity and other factors, this becomes difficult to calculate. The importance of the science is evident though, when we think of the huge bomb-carrying rockets being developed by the U.S. and the U.S.S.R. — the so-called *ballistic missiles*, which are simply "thrown" by rocket, as opposed to *guided missiles*, which are guided by radio waves.

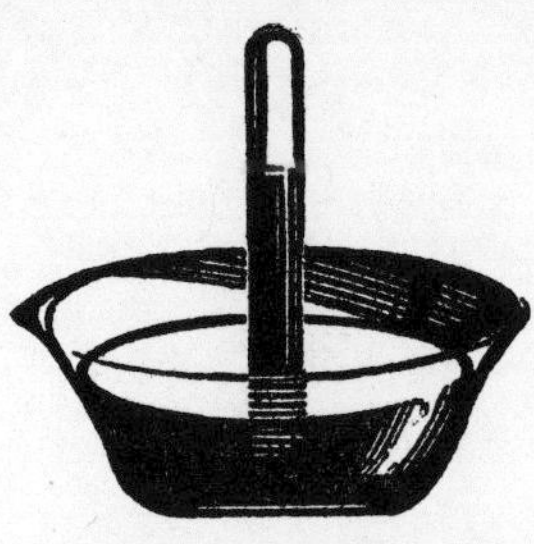

Barometer

IF YOU SUCK at a straw inserted in water, the water level rises within the straw until it reaches the top. If the straw were over 33 feet long, however, and pointed straight up, no amount of suction by mouth or by any form of mechanical pump could pull the water to the top of the straw. The Italian physicist Galileo Galilei pondered over this failure of suction but did not come to the correct conclusion.

Galileo's student Evangelista Torricelli thought, however, that what made the water move upward was not the pull of suction, but the push of air pressure against the water level in the well or container from which the water was being sucked or pumped. When a column of water reaches such a height that the pressure of its weight on the water level of the well is equal to the pressure of the atmosphere's weight, it can rise no higher.

Torricelli tested this in 1643 by using mercury, which is 13.5 times denser than water. He reasoned that a 2½-foot column of mercury would exert as much pressure as a 33-foot column of water, and that should be all the mercury the atmosphere would support. Torricelli filled a 3-foot tube with mercury, placed his thumb over the open end, and upended it into a dish of mercury.

When he took away his thumb, the mercury dropped until the column stood just 30 inches above the mercury level in the dish and there it continued to stay. Air pressure held it that high and no higher. (The 6 inches above the mercury column was a *vacuum*—from the Latin "vacuus," meaning "empty"—the first decent vacuum produced by man. A vacuum produced in that manner is still called a *Torricellian vacuum.*)

In 1648, the French mathematician Blaise Pascal had such a mercury tube carried up a mountainside. As altitude increased, there was less and less air overhead. What was left had less weight and exerted less pressure than air at sea level, so the mercury column should sink lower, and it did.

Such a mercury column is still used today to measure air pressure. It is called a *barometer*, from the Greek "baros" (weight or heaviness) and "metron" (measure). It "measures the heaviness" of air.

Base

THE ASHES of plants yield substances with alkaline properties (see POTASSIUM) which are capable of counteracting the properties of acids. Both acids and alkalis, if strong, can be corrosive and dangerous. However, a strong acid added to a strong alkali results in a mixture that may be very mild; in one case, at least, it will be nothing more than salt water. A substance neither acid nor alkali is *neutral*, from the Latin "ne" (not) and "uter" (either); it is "not either" (neither) one or the other.

The alkaline substances obtained from plant ashes can be made even more alkaline by heating them strongly. Part of the ash turns into vapor and disappears, and this is carbon dioxide. Part remains behind, this being sodium or potassium oxide which, on addition of water, becomes sodium hydroxide (*caustic soda*) or potassium hydroxide (*caustic potash*). The word *caustic* comes from the Greek "kaustikos" which in turn comes from "kaiein" (to burn) — which is a good description of what happens if either compound comes in contact with your skin.

To the early chemists, the part that remained behind after heating naturally seemed the steadier and firmer part of the original ash. So it was called the *base* from the Greek "basis" (pedestal). It formed the pedestal, in other words, upon which the rest of the compound could be built.

Of course, *base* quickly came to mean any compound that could neutralize an acid and this led to a contradiction. Ammonia, for instance, could neutralize acids and yet it was a gas that was given off as vapor on heating. For this reason, ordinary bases were called *fixed bases* ("fixed" in the sense of "tied down") from the Latin "figere" (to fasten), while ammonia was called a *volatile base* from the Latin "volare" (to fly) and "volatilis" (flying from). After all, a gas will fly away, and now we have a "flying pedestal" which is not the most sensible notion possible.

However, modern chemists don't worry about it. A base is any compound which will neutralize an acid, whether the compound be solid, liquid or gas. A strongly basic substance is still called an *alkali* but there is no special name for strongly acidic substances.

Benzene

THERE IS A tree in Indonesia called *benzoin* or *benjoin.* It comes from the Arabic phrase "luban jawi," meaning "incense of Java." A resinous material obtained from incisions in the bark of this tree is *gum benzoin.* An acid easily obtained from this resin is called *benzoic acid.*

In 1834, the German chemist Eilhart Mitscherlich converted benzoic acid into a hydrocarbon (a compound with a molecule made up of only carbon and hydrogen atoms) and named it "benzin." Another German chemist, Justus Liebig, objected that the "-in" ending was being used for compounds that contained nitrogen, which this one did not. He suggested the name "benzol," the "-ol" ending signifying "öl" (German for "oil").

Liebig was a most influential chemist and his suggestion holds good to this day in Germany, but with all respect to Liebig, it is still a bad name. The ending "-ol" is used by chemists for alcohols, which benzol is not. In England, France, and America, therefore, the compound is known as *benzene,* which is best, since "-ene" is an ending reserved for certain hydrocarbons.

But benzene had actually been discovered before the time of Mitscherlich. In 1825, the English electrochemist Michael Faraday isolated some out of an oily residue obtained from illuminating gas. He called it *carburetted hydrogen.* However, in 1837, the French chemist Auguste Laurent, leaving the Germans to fight over "benzin" and "benzol," suggested the name "pheno" from the Greek "phainein" (to shine), in recognition of the fact that the compound was first found in something that shone, i.e., illuminating gas.

This did not catch on for benzene itself, but when a benzene molecule is attached to other atom combinations, the benzene portion of the over-all molecule is referred to now as the *phenyl group,* so Laurent half won. Furthermore, a benzene molecule to which a hydroxyl group (one made up of a hydrogen and an oxygen atom) is attached, is called *phenol.* (Here the "-ol" ending is justified, Phenol is a kind of alcohol.)

Bile

THE JUICE formed by the liver may be called either bile or gall. *Bile* comes from the Latin word "bilis," which is what the Romans called the juice, but *gall* is Anglo-Saxon in origin. Both words are used in combination with others, and the choice of which to use depends on the derivation of the other word.

For instance, the juice is stored in a small pear-shaped bag called a *gall bladder* (never a "bile bladder," since *bladder*, like *gall*, is of Anglo-Saxon derivation), and is led into the intestines by means of a *bile duct* (never a "gall duct," since *duct* is from the Latin "ducere" meaning "to lead").

Sometimes material dissolved in the bile settles out in little crystals which may collect into hard masses. If these get stuck in the bile duct and form an obstruction, the result is pain and possible surgery. Such masses are called *gallstones* (never "bilestones" since *stone* is from the Anglo-Saxon). On the other hand, the Latin word for a small stone is "calculus" (plural, "calculi") and this name may be used for gallstones. But, because of the change to Latin, they are never called "gall calculi," but always *biliary calculi.*

When gallstones obstruct the way to the intestines, the bile being formed by the liver backs up and is forced into the blood stream. The material in bile consists in part of strongly colored compounds called *bile pigments* (never "gall pigments" because *pigment* is from the Latin "pingere," meaning "to paint," and "pigmentum," a paint). In general, these pigments are dark green or brownish red. In the blood, where the color is mixed with the red of blood and is seen through the light yellow of the skin and its underlying fat, the pigments give the complexion a sickly green-yellow appearance.

For that reason, any disorder involving bile in the blood is called *jaundice* (from the French "jaune," meaning "yellow") and when the disease is caused by a gallstone obstruction, it is called *obstructive jaundice.*

Bromine

Few chemical substances are named as the result of the sense of smell; very few in comparison with the names resulting from the sense of color vision. However, there are some familiar cases.

For instance, in 1824, a young French chemist Antoine-Jérôme Balard was studying some of the crystalline material obtained from the brine of a salt marsh. He noticed that when certain chemicals were added to a solution of this brine, a brownish color appeared. He investigated this and found a new element, one of the few that is liquid at ordinary temperature. It is a deep red in color and has a strong odor, something like that of chlorine or iodine but stronger.

Balard suggested it be called "muride," from the Latin "muria" (brine), but this didn't catch on. Instead, the element was named *bromine* from the Greek "bromos" (stench), in honor of its odor. (Actually, there are chemical compounds that smell infinitely worse than bromine, so it has always seemed unfair to me to pick on the poor element in this way.)

In 1839, to take another case, the German chemist Christian F. Schönbein discovered a gas which turned out to be a variety of oxygen. Whereas ordinary oxygen had a molecule made up of two oxygen atoms, the new gas had one made up of three oxygen atoms. The three-atom gas has a marked odor (something like that of weak bromine) and Schönbein named it *ozone*, from the Greek verb "ozein," meaning "to smell."

But prior to both those discoveries, the same situation had arisen with the same result. The British chemist Smithson Tennant, in 1803, found, when working with crude platinum, that after the platinum had been dissolved in a mixture of acids, a black metallic powder remained behind. Investigating that powder, he found two new elements. One of them formed a compound with oxygen that even in small quantities had a strong smell like that of chlorine. So he named that element *osmium*, from the Greek noun "osme," meaning "a smell."

Caffeine

There is a plant, native to Ethiopia, the seeds of which when roasted, ground, and soaked in boiling water, make an immensely popular drink. Its name is derived, according to one theory, from the Ethiopian province in which it may have been first grown, the province of Kaffa.

The drink spread from Ethiopia to Arabia where it quickly grew popular. Islam forbade the drinking of intoxicating liquors and while this was all very well, some sort of stimulating drink becomes popular with people despite all sorts of laws. The Ethiopian seed plus water offered a drink that, though not intoxicating, was stimulating enough to be an acceptable substitute for wine to the Arabs. So according to a second theory, it derives its name from an Arabic word for wine, "qahwah."

From Arabia, the drink spread through Europe in the 1600's, and there the French called it *café* and the English, *coffee*. Little places which would specialize in selling coffee and other things to eat or drink on the side became so popular that *café* and *cafeteria* (the Spanish word for "coffeehouse") are English words for kinds of restaurants now.

The German chemist F. Runge, in 1820, isolated from coffee seeds (or beans, as they are usually inaccurately called because of their shape) an alkaloid which was the stimulating ingredient of coffee. He named it, very naturally, *caffeine*.

The Chinese flavored their boiled water with the leaves of a shrub they called "ch'a" or "tse" or "te" (depending on the province). This also swept Europe, later than coffee did, and won its outstanding victory in England where it completely replaced coffee. The Chinese drink is called *chai* in Russian, *Thee* or *Tee* in German, *thé* in French, *Thee* in Dutch, and *tea* in English. Tea leaves, too, contain caffeine, but when first isolated the caffeine from tea was thought to be a distinct substance and named *theine*.

Caffeine is also found in seeds of a Brazilian shrub, named the guarana, used to make stimulating drinks, and the caffeine therein was named *guaranine*.

Calciferol

INFANTS SOMETIMES develop bones that are soft and deform easily under weight or under muscle-pull, so that bowlegs or curved spines or misshapen skulls result. This is called *rickets* or *rachitis*, possibly from the Greek "rhachis" (spine). A misshapen, soft skeleton results in weakness, and the common adjective *rickety*, meaning "ready to fall," comes from the name of this disease.

In 1918, the English physiologist Edward Mellanby discovered a substance that, when fed to infants, seemed to prevent the development of rickets. This *antirachitic factor* was named *vitamin D* (see VITAMIN), in 1922, by the American biochemist Elmer Verner McCollum.

By 1935, the molecular structure of vitamin D was worked out and it was found to resemble that of certain sterols (see CHOLESTEROL). In fact if these sterols are exposed to ultraviolet light, the necessary change into vitamin D takes place. The Latin word for "to shed light upon" is "irradiare" from "radius" (ray). "Radiare" means "to shed light" and the prefix "in-" which means "on" has become here "ir-." And so foods containing the necessary sterols may be *irradiated* until they contain the vitamin instead. The human skin contains some of the necessary sterols, and sunlight contains enough ultraviolet to irradiate us effectively. For that reason, vitamin D is sometimes called the sunshine vitamin.

The action of vitamin D is to encourage, in some way, the transfer of the calcium ion floating freely in blood to the bone where it is firmly bound into a crystal. The chemical name for the vitamin is therefore *calciferol*. The Latin "ferre" means "to bear," so vitamin D is the "calcium bearer."

Different varieties of the vitamin, each about equally effective, may be formed from different sterols. From ergosterol (a sterol occuring in a type of fungus called ergot), *ergocalciferol* (vitamin D_2) is formed; and from 7-dehydrocholesterol (the sterol in our skins), *cholecalciferol* (vitamin D_3) is formed. (Vitamin D_1 does not exist. The name was originally given to a preparation that turned out to be a mixture of different substances.)

Calcium

The Latin word for "stone" is "calx" (genitive form, "calcis") and was applied particularly to a certain common type of stone which, in English, we call *chalk* (from "calx," you see). In its most beautiful form, this stone is called *marble,* from the Greek "marmaros" which means "sparkling stone." And it is true that much of the beauty of marble comes from a sort of fine-grained sparkle.

The Anglo-Saxon name for the material, by the way, is limestone. When it is heated, it gives off carbon dioxide, and what is left is called lime. However, when, in 1808, the British chemist Sir Humphry Davy first prepared a new element out of lime, he went back to the Latin for the name. He added the "-ium" suffix common for metals and named the new element *calcium.*

(Calcium is not the only element to get its name from a common stone. The Latin word for flint (itself of Anglo-Saxon derivation), a rock even more common than chalk, is "silex" (genitive form, "silicis"). Consequently, the early chemists called flint and similar rocks *silica.* Then when, in 1824, the Swedish chemist Jöns J. Berzelius found a new element in silica, he simply added the "-on" nonmetal suffix and the result was *silicon.*)

But stones achieve fame in another direction also. The Latin language generally indicates smallness by adding "-ul" to words. Since "calcis" means "stone," "calculus" is a "small stone" or "pebble."

Pebbles were useful in solving arithmetical problems. The earliest mechanical device for solving such problems was the *abacus* (a Latin word from the Greek "abax" (genitive, "abakos"), meaning a board on which one solved problems). This consisted of pebbles strung on wires, or placed in grooves, in a wooden framework. By moving the pebbles properly, problems in arithmetic could be solved.

Hence, to work out an arithmetical problem is to *calculate.* Furthermore, when, in 1666, the British mathematician Isaac Newton devised new mathematical methods for solving problems that could not be handled by the old methods, the new methods were eventually called *calculus.*

Calendar

THE EARLIEST way of measuring periods of time longer than the day involved the moon. It is impossible to avoid noticing how night by night the moon changes from a crescent to a full circle and then back to a crescent. (The very word *crescent* comes from the Latin "crescere," meaning "to increase." The final shrinking form of the moon should, strictly speaking, be called a "decrescent" but it isn't.)

To primitive peoples it would seem that periodically a new moon was created and we still call the early crescent moon a *new moon.* It was convenient to measure time by the number of new moons that had appeared in the sky. The period from one new moon to the next is just about 29½ days and that period is the original *month.* In English, at least, the connection between *month* and *moon* is obvious.

The early Romans kept time strictly by the moon. They even had a ritual whereby the high priest would watch every month for the first appearance of the new moon's crescent. Once it appeared, he would officially proclaim the beginning of the new month. The first day of the month was therefore called the *calends,* from the Latin "calare" (to proclaim). It was easy for this word to spread its meaning over the entire month and now we call a table of months, or a system for telling time by months, the *calendar.*

For periods less than a month, it is convenient to use the half-moon. From new moon to half-moon is just under 7½ days. The same period of time exists from half-moon to full moon, from full moon back to half-moon, and from half-moon back to crescent.

Consequently, the Babylonians divided the month into seven-day periods. The Jews picked up the habit during the Babylonian Captivity and spread it to the rest of the world via Christianity. The English word *week* for this seven-day period is from the Anglo-Saxon but may hark back to an old Gothic word meaning "change" (of the moon, you see).

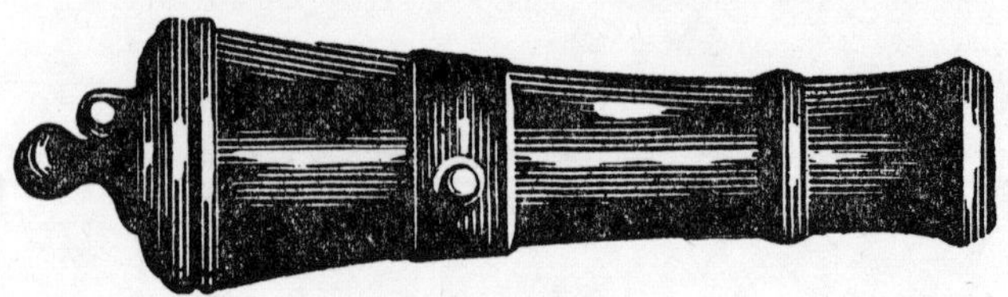

Calorie

BEFORE THE 1850's, chemists and physicists thought that heat was some kind of separate substance in matter that would flow from a hotter body to a colder one. It could flow into water to make it steam, and out of burning coal to heat the air. This substance was called *caloric*, from the Latin word "calor," meaning "heat."

Caloric ran into trouble, however, when the American-born scientist Benjamin Thompson (he was a Tory during the Revolution and had to emigrate, eventually getting a title and becoming Count Rumford) noticed in 1798 that drilling brass cannon produced quantities of heat. He reasoned that both drill and cannon were cold to begin with and contained little caloric. Where did the unlimited supply of caloric come from once drilling began? The caloric theory was never quite the same again.

In 1857, the German physicist Rudolf J. E. Clausius advanced the theory that heat was not a material substance, but energy — the energy of vibrating molecules. The heat of drilling thus came from the mechanical energy involved in forcing the drill through the brass. This view has been accepted ever since.

But an echo of the dead caloric still remains. To measure the quantity of heat, physicists decided to call the heat required to raise one gram of water from a temperature of 14.5°C. to 15.5°C. one *gram-calorie* or, simply, one *calorie*. (Since heat is a form of energy, it can also be measured in ergs and joules (see ENERGY). One calorie is equal to 4.185 joules or to 41,850,000 ergs.)

A larger unit is frequently used. It is the *kilogram-calorie*, or *kilocalorie*, which is equal to 1000 calories. Unfortunately, it has become too customary to call the kilocalorie simply *Calorie* (with a capital C). The capital cannot be heard by the ear and this is a source of great confusion. People who talk of the "calorie content" of various foods are almost always talking about the kilocalorie content.

Capillarity

MOLECULES ATTRACT neighboring molecules. They will attract others like themselves so that a piece of steel, for instance, doesn't fall apart into dust even under thunderous blows, and a drop of water may hang from a faucet for quite a while before dropping off. This is *cohesion*, from the Latin "co-" (together) and "haerere" (to stick). A substance will "stick together."

Molecules will also be attracted to other molecules unlike themselves. Paint will stick to wood, mortar to brick, glue to almost anything. This is *adhesion*, the Latin prefix "ad-" meaning "to." Paint "sticks to" wood.

Water in a tube will cohere. The water molecules near the tube itself will adhere to the glass or cellulose of the tube. The adherence is actually stronger than the coherence, so that where water meets tube, the level curves up in order that as much water as possible be next the tube. In a narrow tube, say a half-inch in diameter, the upward curve of the water level forms a crescent that is called a *meniscus*, from the Greek "meniskos" meaning "a little moon." (If the tube were made of wax, water cohesion would be stronger than the water-wax adhesion and the meniscus would curve downward all around. The same is true for mercury in a glass tube, since mercury cohesion is unusually high for a liquid.)

If tubes are fine enough, the adhesive forces are enough to pull the water level a considerable way upward against gravity. Stick a blotter into water, for instance, and the water will soak upward through the fine spaces between the matted cellulose fibers of the blotter. Water will also soak upward through the very fine tubes leading up tree trunks, and this is one of the mechanisms by which trees lift water hundreds of feet without a pumping mechanism like the animal heart. Because this works best in tubes that are as fine as hairs, or finer, this lifting of fluids is termed capillary action or *capillarity* from the Latin "capillus" (hair).

Carcinoma

PERHAPS THE most frightening word in the medical vocabulary is *cancer* and yet in Latin it is an innocent word meaning "crab." We still use the word in its innocent sense as a sign of the zodiac (Cancer, the Crab; Tropic of Cancer. See SOLSTICE). The reason for the name is uncertain. It dates back to ancient times but no one suggests that people connected the disease with the influence of the constellation (see INFLUENZA). Perhaps it's just that cancer clings to what it attacks as though by crabs' claws, or that it sometimes seems to take on a crab-like shape.

The word *tumor* is not synonymous with *cancer* although the average man sometimes uses it in that way. A tumor is any abnormal growth on the body, from the Latin "tumere" (to swell). A tumor that is of limited growth, like a wart, may be unsightly, but it is no danger, so it is a *benign tumor* (from the Latin "bene," meaning "good," and "genus," meaning "sort"; a "good sort of tumor").

If, on the other hand, the tumor is of the type that grows without limit until it kills, it is a *malignant tumor*. Since the Latin "malus" is "bad," it is a "bad sort of tumor," and it is the malignant tumor that is cancer.

The most common types of cancer are those which strike the parts of the body that face the outside world, the skin or the linings of the alimentary canal. These are *carcinomas*, from the Greek "karkinos" (crab), so that, except for the switch in language, it is just another way of saying "cancer."

The suffix "-oma" is used in modern medical terminology to mean a tumor — usually, but not always, a cancerous one. A cancer within the body attacking the connective tissue is a *sarcoma* from the Greek "sarx" (flesh). An *adenoma* is a tumor of glandular tissue and a *hepatoma* is cancer of the liver. (The Greek "aden" means "gland" and "hepar" means "liver.")

Leukemia is an exception. It is a condition in which the white cells of the blood multiply in cancerous fashion. The ending in this case is from the Greek "haima" (blood), while "leukos" means "white" and refers to the white cells (see HEMOGLOBIN), so it means "white cells in the blood."

Carnivore

Diet is not necessarily a matter of choice. A cat, for instance, has teeth suitable for tearing food, but not for grinding it. It lacks our copious saliva with which to moisten food. It has a relatively short digestive tract so that it can't indulge in prolonged handling of the food it eats. For all these reasons, the main items in its diet must be food that can be quickly digested even when swallowed in chunks. So it lives on other animals and is a "meat-eater" or *carnivore*, from the Latin "caro" (genitive, "carnis" meaning "meat") and "vorare," "to eat" or "devour."

Although there are carnivorous birds, fish, insects, and even plants, the most familiar carnivores are mammals and many of these are included in a mammalian order entitled *Carnivora*.

Only one other mammalian order (the one including shrews and moles) is named after its feeding habits and it involves meat-eaters, too. The meat, however, is specialized, consisting mostly of insects. The order is therefore called *Insectivora*, or "insect-eaters."

There are also a number of familiar animals with teeth designed to grind coarse food and with long complicated digestive systems to handle that food. (Coarse food like grass may be hard to digest, but at least it doesn't have to be chased and doesn't fight back.) The "grass-eaters," such as cattle, sheep, and horses, are *herbivores*, from the Latin "herba" (grass). These are spread over a number of mammalian orders.

Then there are specialized herbivores. Although many bats are insectivores, some of the large ones, notably the *flying foxes* (so called because they have heads like those of small foxes), are *fruit bats* (so called because they live on fruits). "Fruit-eaters" are *frugivores*, from the Latin "frux" (fruit).

Of course, there are creatures who can eat almost anything, plant or animal. Such animals are generally successful in life and include bears, pigs, rats, crows, and, most particularly, man. These are the *omnivores*, from the Latin "omnis" (all); they "eat all."

Carotid

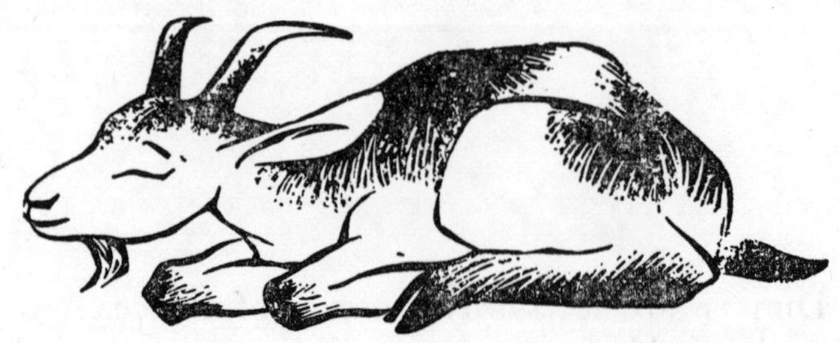

THE NAMES OF arteries and veins usually come from the Greek or Latin names of the organs they serve. For instance, the *hepatic artery* leads blood to the liver, the Greek word for "liver" being "hepar." In the same way the *renal artery* leads blood to the kidney and the *pulmonary artery* leads it to the lungs. The Latin word for "kidneys" is "renes" and for "lung" is "pulmo" (genitive, "pulmonis").

There are exceptions, though. There is the *jugular vein*, which passes through the neck. It is named not after the organ it serves, but after the part of the body through which it passes. The Latin "jugulum" means "collarbone" or, more generally, "throat."

The arteries that serve the heart surround and encircle it like a crown. The Latin word for "crown" is "corona" so the arteries to the heart are the *coronary arteries.* Any interference with those arteries means quick and painful trouble, since the heart can't stop working and must have nourishment. If a clot forms in the arteries, cutting off some of the blood supply, the result is one form of "heart attack." The Greek word for a "clot" is "thrombos" so such a heart attack is called a *coronary thrombosis.* For this reason, the word *coronary* by itself has become a slang term for "heart attack."

The brain, like the heart, is sensitive to lack of blood, but this shows up in another way. If the supply of blood to the brain is cut down, the result is not pain, then death; but sleep, then death. Sideshow performers in ancient Greece, for instance, used to amaze their audiences by pressing a spot on a goat's neck (pinching off the artery leading blood to the brain) and causing it to go to sleep. Releasing the pressure would allow it to wake again. The Greek word for "stupefy" is "karoun" and the arteries leading to the brain are therefore called the *carotid arteries.*

Catalysis

THE ANCIENT philosophers speculated on the possible existence of some substance which, by its mere presence and without being used up in the process, could change base metals into gold. This was called the *philosophers' stone* and, of course, does not exist. However, other and infinitely more valuable "philosophers' stones" have been discovered.

As early as 1750, sulfuric acid was being formed in large quantity from sulfur dioxide and water by the use of nitrogen oxides which, somehow, made the appropriate chemical changes take place without being used up in the process. (And sulfuric acid is incomparably more valuable, though much less costly, than gold.)

In 1823, the German chemist Johann Wolfgang Döbereiner invented a kind of lighter in which a jet of hydrogen was directed on a bit of platinum. The hydrogen at once caught fire because the platinum brought about its combination with the oxygen in the air without itself being used up.

Ordinary substances, too — simple acids, for instance — could bring about changes in this way, as in breaking down starch to sugar.

In 1836, the Swedish chemist Jöns Jakob Berzelius reviewed the whole subject. Though he could not explain it, he suggested a name. He called the "philosophers' stone" process *catalysis* from the Greek "katalysis" meaning "dissolution" or "destruction" ("kata", "down", and "lysis," "breaking"). The substance that brought about the catalysis was the *catalyst* which, by its mere presence, destroyed sulfur dioxide, or hydrogen or starch.

We have since learned that it is more than a case of "mere presence"; that catalysts take part in the reactions they catalyze, but are reformed before the reaction is over, so that they seem not to have changed. Chemical industry, today, depends almost entirely on the use of appropriate catalysts and so does all living tissue, including our own (see ENZYME).

Celluloid

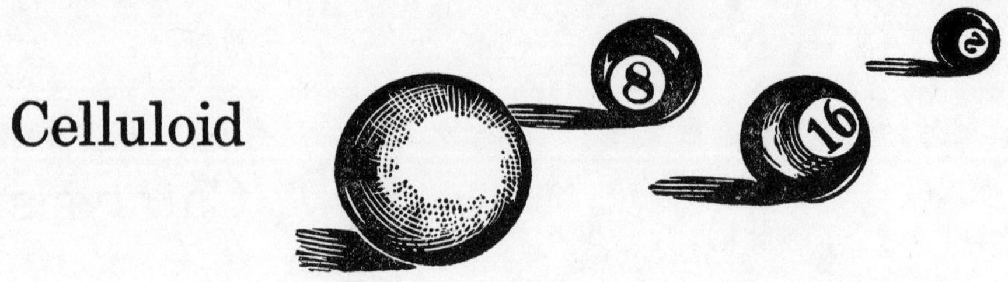

ALTHOUGH POOLROOMS are frequently pictured as places where idlers waste their time and from which little good can be expected, the game of billiards was nevertheless the cause of an important chemical discovery. The billiard ball, it seems, must be hard, elastic, uniform in composition, and take a high polish. The ideal material for the purpose is ivory, which is the substance forming the tusks of elephants. These are a pair of enormously large teeth, and the word *tusk*, in fact, comes from the same Anglo-Saxon root, "tux," from which "tooth" itself is derived.

Now there are things in this world easier to procure than elephants' teeth and in the 1860's a prize was offered for the discovery of an adequate ivory-substitute for billiard balls. An Englishman A. Parkes had already discovered that if camphor is added to pyroxylin (see RAYON), the mixture becomes plastic; that is, it can be molded into any desired shape. Camphor is an example therefore of a *plasticizer*.

The American inventor John Wesley Hyatt studied the camphor-pyroxylin product (which the British called *Xylonite*, from "pyroxylin") and worked out practical mechanical devices for turning out Xylonite billiard balls, winning the prize in 1870. Hyatt called the material *celluloid*, since pyroxylin is derived from cellulose and the "-oid" suffix is from the Greek "-oeides" (having the form of). And thus billiards was responsible for the commercialization of the first artificial plastic.

The biggest shortcoming of celluloid was its inflammability. What was really needed for most purposes was a hard plastic that was also stable. The first thermosetting plastic (see POLYMER), one which after all these years can scarcely be improved on for hardness and stability, was formed by the polymerization of phenol and formaldehyde. This was accomplished in 1906 by the Belgian-born American chemist Leo Hendrik Baekeland who, more self-assertive than Hyatt, named his product after himself. It is *Bakelite*.

Centigrade

IN GENERAL, substances expand slightly when heated, and shrink when cooled. This fact gave mankind its first tool for measuring temperature accurately — the mercury thermometer. *Thermometer* comes from the Greek words "therme" (heat) and "metron" (measure); it is an instrument "to measure heat."

The mercury thermometer was invented in 1714 by the German physicist Gabriel Daniel Fahrenheit who filled a hollow bulb with mercury and allowed it to expand, when heated, up a very fine enclosed and evacuated tube. The amount by which the mercury thread crept up the tube was proportional to the temperature. (The glass also expanded but nowhere nearly as much.)

Fahrenheit placed the mercury-filled bulb in a mixture of equal parts of salt and snow at the melting point and marked the height of the mercury column as 0. He next let it warm to the temperature of the human body and marked the new height as 100. (The man he used must have been slightly feverish or else Fahrenheit adjusted the level to allow the freezing point and boiling point of water to come out at whole values.) By drawing a hundred equal divisions between the two marks, he invented the *Fahrenheit scale*. On this scale the melting point of pure ice is 32 degrees and that of the boiling point of pure water is 212 degrees. The word *degree* comes from the Latin "de-" (down) and "gradus" (step). In marking off small divisions from 100 to 0, you go "down steps."

In 1742, the Swedish astronomer Anders Celsius suggested that the temperature of melting ice be set at 100 degrees and that of boiling water at 0 degrees. (The 0 and 100 were later reversed.) This hundred-degree interval gave the new scale the name *centigrade* from the Latin "centum" (hundred) and "gradus" (step). It is the scale of "a hundred steps" from melting to boiling. It is also called the *Celsius scale* after the inventor.

The Fahrenheit scale is commonly in use among the general public in England and America, but it is the Celsius scale that is used in other civilized nations and among scientists everywhere, including those in England and America.

Centipede

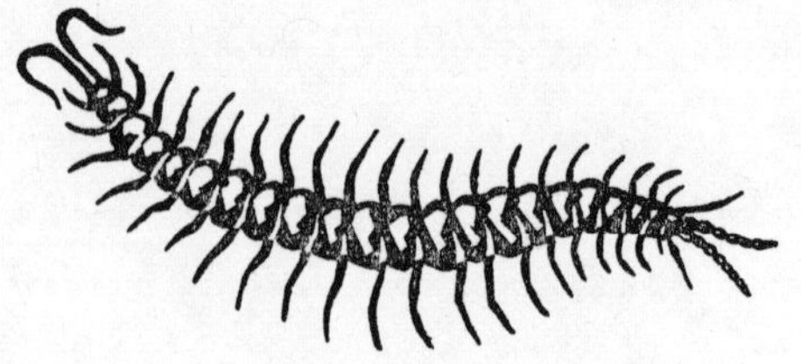

MAN MUST early have realized that one outstanding difference between himself and the other animals was that he walked on two legs, the others on four. The various mammals and reptiles (but especially mammals) were therefore lumped as *quadrupeds*, from the Latin "quattuor" (four) and "pes" (foot) — the "four-footed ones."

Birds also had four limbs, but two were used as wings and only two as legs. They were *bipeds* from the Latin "bis" (twice) — the "two-footed ones." Man, of course (also the kangaroo, various jumping rats, and running lizards, to say nothing of certain dinosaurs), is a biped.

The insects, one and all, are blessed with six legs. This is taken notice of, and the class name Insecta is sometimes referred to (but much less frequently) as *Hexapoda*, from the Greek "hex" (six) — the "six-footed ones." (The Greek word for "foot" is "pous," genitive, "podos.")

Spiders are not insects despite the popular misconception. Spiders lack wings, have two main body-segments rather than three and, in common with their relatives the scorpions and king crabs, have eight legs. Nevertheless, the name *Octopoda* is reserved for an order of sea creatures that don't have feet at all but do have eight tentacles. (The word *tentacle* comes from the Latin "tentere," meaning "to touch.") The most common of these is the *octopus*. In both Latin and Greek "octo" means "eight," so the octopus is the "eight-footed one."

And there are crawling creatures with many more than eight legs. They are the *centipedes* and the *millipedes* who were once grouped under the name *Myriapoda*. The Latin "centum" means "a hundred" and "mille" means "a thousand" while the Greek "myrios" means "ten thousand" so you get the idea. Both centipedes and millipedes are composed of many body segments. From each segment in the centipede are produced one pair of legs; from each segment in the millipede, two pairs of legs. The millipedes are therefore put in the class *Diplopoda*, from the Greek "diploos" (double), hence the "double-footed ones."

Cerebrum

Until modern times, the brain was little regarded (see pituitary). In fact, to the ancients, the most remarkable thing about heads were horns, and if the brain got a name it was apt to come from the word for horn, since the brain was something that was in the same part of the animal as its horns (if it had any) were.

The word *horn* itself is related to the Latin word for it, "cornu." (We keep the older pronunciation, by the way, when we speak of a horny callous on a toe and call it a *corn.*) The German word for "brain" is *Hirn* and the similarity to *horn* is obvious. The similarity of *brain* itself is less obvious and may not exist. An alternative theory is that it comes from the Greek "bregma" meaning the top portion of the head, or from a common ancestor.

However, we are not through with *horn.* The Greek word for it is "keras" and when the Romans called the chief portion of the brain the *cerebrum* (remember that in Latin, *c* is always pronounced "k") the relationship to *horn* seems clear. Behind and under the cerebrum is a smaller structure called the *cerebellum.* This is a Latin diminutive of *cerebrum* so that it means literally "little brain." Both *cerebrum* and *cerebellum* have come into the English language unchanged.

A word sometimes used to mean the whole brain is *encephalon* which comes from the Greek "en-" (in) and "kephale" (head), so that it simply means that which is "in the head." The word is better known in connection with virus infections leading to inflammations of the brain. This is called *encephalitis*, the Greek suffix "-itis" being commonly used in medical terms to mean "inflammation."

A particular variety of encephalitis, which brings on a *coma* (from the Greek "koma," meaning "sleep") is *encephalitis lethargica* ("lethargica" is the feminine form of the Latin word meaning "drowsy," which in turn comes from the Greek "lethe," meaning "forgetfulness," and "argos," meaning "idle"). In common English, the disease is called sleeping sickness.

Chiroptera

WHEN CREATURES have wings, this is a notable thing and the Greek "pteron" (feather) and "pteryx" (wing) are apt to show up in their scientific names. This is true of reptiles (see PTERODACTYL) and insects (see LEPIDOPTERA). It is also true of mammals.

The one group of flying mammals includes, of course, the *bats*. This word can be traced back to an old Scandinavian word meaning "to flutter," and the flight of the bat is indeed much more a fluttery thing than is the flight of birds. We still have this old meaning of *bat* when we "bat our eyes."

This notion is used most directly in connection with the additional fact that the most common English bats are just about the size and general appearance of a mouse (if you ignore the wings), and a very descriptive alternate name for the creature is *flittermouse*.

It is interesting that the animal kingdom has developed wings four different times and each time in a different style. First, the insects developed ordinary membranes with no bony stiffening. Then, reptiles such as the pterodactyl developed membranous wings with a single finger bone as stiffening, the others remaining outside as separate claws. Then, birds developed feather-covered wings with all finger bones, fused together, as stiffening. Finally, the bat has a naked wing strung along four enormously lengthened fingers, so that only its wing of all those developed can flex like a hand. The order that includes the bats has therefore been given the name of *Chiroptera*, from the Greek "cheir" (hand) and "pteryx" (wing). They are the "hand-winged" creatures.

In the case of birds, however, there is little mention of wings in the scientific names. Two examples come to mind, though. There is an extinct bird, halfway from the lizard, that is the *archeopteryx*, from the Greek "archaios" (ancient); it is the "ancient winged" creature. And there is the New Zealand *kiwi* (so called from the sound it utters). It is the one bird that is just about completely without wings, so it is called *apteryx* or "no-wings" since the Greek "a-" means "no."

Chlorine

THE GERMAN CHEMIST Johann Rudolf Glauber treated ordinary salt with sulfuric acid in 1658 and got a solution which gave off a choking vapor or "spirit" (see GAS). He called the new substance *spirit of salt.*

The substance had acid properties in solution and since it was prepared from salt, and salt was most easily prepared from sea water, the new substance was eventually named *marine acid* or *muriatic acid.* (The Latin word "mare" means "sea," while "muria" means "brine.")

Now the so-called muriatic acid happens to be the most common example of an acid that does not contain oxygen. But in the late 1700's, there was a theory that all acids must contain oxygen and, in fact, that is how oxygen received its name (see OXYGEN). It seemed certain, therefore, that the muriatic acid molecule must be made up of atoms of oxygen and of some unknown element which was named *murium.*

In 1774, the Swedish chemist Karl Wilhelm Scheele treated muriatic acid with manganese dioxide and obtained a greenish, chemically active gas with an unpleasant odor. He didn't realize he had a new element because the false theory about acids misled him. He considered it just muriatic acid with additional oxygen (from the manganese dioxide). In fact, the French chemist Count Claude Louis de Berthollet suggested in 1785 that the greenish gas be called *oxymuriatic acid*, and others suggested the name *murium oxide.*

It wasn't until 1810 that Davy (see POTASSIUM) first realized the truth and got credit for finding the new element. He could find no oxygen in "murium oxide" and came to the daring conclusion that there was no oxygen in muriatic acid in the first place. He decided that the greenish gas was a new element, and he overthrew all the previous misleading names and started fresh by calling it *chlorine*, from the Greek "chloros," meaning "green."

As for muriatic acid, its molecule is made up of one atom of hydrogen and one of chlorine. Its modern name is *hydrochloric acid.*

Chloroform

SCIENTISTS OBTAIN new chemicals in odd places. For instance, the English naturalist John Ray, back in the seventeenth century, cooked up a batch of red ants and obtained a liquid which he (or someone after him) called *formic acid*, from the Latin "formica," meaning "ant." At that, it was a good name, because the pain of the bite of the red ant is due to a small quantity of formic acid which it injects into the flesh as it bites. Formic acid is strong enough as an acid to sting badly. It is also, by the way, supposed to occur in nettles, which also explains a good deal.

The molecule of formic acid is made up of a single carbon atom to which are attached (1) a hydrogen atom, (2) an oxygen atom and (3) the hydrogen-oxygen combination called a hydroxyl group. Minor changes in these attachments result in other compounds which often retain the stem "form" in their names. (It often happens in chemical terminology that one substance names another until the original derivation is lost sight of.)

For instance, replace the hydroxyl group with a hydrogen atom and the resulting molecule contains the aldehyde combination of elements (see KETONE). This particular aldehyde would, naturally, be *formaldehyde.* A 40 per cent solution of formaldehyde (which is a gas in the pure state) in water is *formalin* and it is formalin which is used to preserve tissues against decay in zoology and anatomy laboratories. Students of either subject are well acquainted with the odor.

Again, if the hydroxyl group is replaced by a chlorine atom and the oxygen atom is replaced by two more chlorine atoms, the result is a kind of chlorinated formic acid. The name of the compound is rather in the nature of an abbreviation of this, for it is *chloroform* — a familiar name to most people, yet only a chemist would see the connection with ants.

Naturally, if bromine or iodine atoms are present instead of chlorine (and they can be), the results are, respectively, *bromoform* and *iodoform.*

Chlorophyll

Usually, when the prefix "chlor-" appears in the name of a chemical substance, it is a pretty good sign that the molecule contains one or more atoms of chlorine (see CHLOROFORM). However, chlorine itself was named for its green color (see CHLORINE) and sometimes the prefix refers only to the color of a substance and not to its chemical make-up.

The most important green in nature is, of course, the green of plants. The substance responsible for this color was first isolated in 1817 by the French chemists Pierre J. Pelletier and Joseph B Caventou, who named it *chlorophyll* from the Greek "chloros" (green) and "phyllon" (leaf). Here the "chloro-" refers only to the color, for there are no chlorine atoms in chlorophyll.

Chlorophyll acts to trap sunlight and convert carbon dioxide of the air into the food that all animals live on, so it is beyond compare the most important colored compound on earth. There are, however, other colored compounds in plants which are infinitely less important and yet which we would hate to have to do without.

For instance, there is a type of colored compound called *anthocyanin* that appears in flowers of many types. The name is derived from the Greek "anthos" (flower) and "kyanos" (blue). From this you would think it appears in blue flowers, and so it does in some, such as the delphinium. However, under the proper circumstances, anthocyanins can be violet or red and are responsible for the color of the violet, dahlia, poppy, and rose.

There are yellow pigments in plants, too. Some are called *flavones* (from the Latin "flavus," meaning "yellow") and some *xanthones* (from the Greek "xanthos," meaning "yellow"). However, the most common yellow-orange pigment in plants is *carotene*, so named because it was first isolated from carrots. It also occurs in many plant and animal fats, including that of man. Some peoples of East Asia have enough carotene in the fatty layers under the skin to cause people to speak of a "Yellow Race."

Cholesterol

In 1769, a French chemist, Poulletier de la Salle, working with gallstones (see BILE), obtained from them a white solid substance with a fatty feel. In 1815, another French chemist, Michel Eugène Chevreul, decided the substance was a kind of fat and named it *cholesterine*. The word came from the Greek "chole" (gall) and "stear" (a hard fat).

In the early days of chemistry, the suffix "-ine" or "-in" was a common one for the organic compounds in living tissue. As more and more came to be known about the chemical constitution of such compounds, it became customary to reserve the "-ine" suffix for those that contained one or more nitrogen atoms in their molecules. Cholesterine did not have such nitrogen atoms. In 1859, moreover, still another French chemist, Pierre E. M. Berthelot, showed that cholesterine did have a hydroxyl group in its molecule, so that it was actually an alcohol (see ALCOHOL). The chemical names of alcohols have an "-ol" suffix, but it was not until 1900 or thereabouts that the switch was made and the compound was finally named *cholesterol*.

Slowly, more facts were learned about its chemical structure. It wasn't until the 1930's that the last details were worked out, but by 1910, chemists were reasonably certain that the carbon atoms of cholesterol were arranged in a number of connected rings to which other carbon atoms (side chains) were attached. Other compounds were known by then which had the same ring system but slight differences in the side chains. By 1911, such compounds were given the general name *sterols*.

Then additional compounds were discovered with the same ring system, but lacking the hydroxyl group that made cholesterol an alcohol (though containing oxygen atoms in other combinations) and therefore not deserving of the "-ol" suffix. A still more general name was therefore invented in 1936: *steroid*, the suffix "-oid" being the usual Greek-derived ending meaning "in the shape of." The ring combination found in all such compounds is now called the *steroid nucleus*.

Chromatography

One of the difficulties in studying the chemistry of living tissue (a science called *biochemistry*, from the Greek "bios," meaning "life," hence "life-chemistry") is that tissues consist of a mixture of a vast number of compounds, groups of which are so similar that they are almost impossible to separate. For instance, there are a number of very similar colored compounds in leaves, and finding out anything about them was next to hopeless unless they could be separated, and separating them was almost hopeless, too.

In 1906, however, the Russian botanist Mikhail Tswett found the answer. He took the mixture of leaf pigments, dissolved it in petroleum ether, and poured the solution through a glass tube packed tightly with powdered limestone. The petroleum ether went through but the pigment clung to the tiny limestone particles and remained behind (see ADSORPTION).

However, as he continued to pour fresh petroleum ether through the column, the pigment was slowly washed downward. Each separate compound in the mixture was washed down at a slightly different rate, depending on how firmly it was attached to the limestone particle and how easily it dissolved in petroleum ether. The result, as one particular pigment moved down quickly, another slowly, and another still more slowly, was that separate bands appeared in the column. Each band contained one particular pigment separated from the rest, and each was its own shade of red, orange, or yellow.

Tswett called this technique *chromatography*, from the Greek "chroma" (color) and "graphein" (to write) because the solution to the mixture problem was "written in color" for everyone to see on the limestone column. It took 25 years for Tswett's discovery to be adopted by biochemists (no one seems to listen to Russian scientists) but today it is one of biochemistry's most powerful tools. Limestone has been replaced by more efficient powders and very often by simply a sheet of absorbent paper. Furthermore, the technique is used mostly for colorless compounds, now, but it is still called chromatography.

Chromium

Some elements are more apt to form colored compounds than are others. A particular case in point is a silvery metal first isolated by the French chemist Nicolas-Louis Vauquelin in 1797. In working with this metal, he obtained several of its compounds in red, yellow, and green. As a result of this, the name suggested for the new element was *chromium*, from the Greek "chroma" (color).

This has had one odd result. Chromium is one of those metals which can be electroplated onto steel. The chromium "skin" not only takes a high and good-looking polish, but it is also resistant to rust so that it protects the steel beneath, which would otherwise quickly rust and crumble. Chromium-plated steel is used commonly in the decorative metalwork on automobiles and it has become customary to speak of such metal as "chrome," so that the Greek word for "color" ends by being applied to a handsome but entirely colorless material.

More appropriately, the word "chrome" also occurs in the names of the colored chromium compounds which are now used as pigments in paint. A compound of chromium and oxygen, for instance (called *chromic oxide*, logically enough), is also called *chrome green* because of its color. If chromium and oxygen are combined with lead in various ways, *lead chromate* is formed which may appear in different colors and is called *chrome red*, *chrome orange*, and *chrome yellow*.

A simliar discovery occurred five years after Vauquelin's. The English chemist Smithson Tennant discovered a new element in crude platinum. This new element was also unusual for the number of differently colored compounds it formed and Tennant named it *iridium* from the Greek "iris" or rainbow (genitive, "iridos"). It is interesting to note that to the Greeks, Iris was the messenger of the gods, and it was only natural to suppose that the rainbow, which seems to stretch between earth and heaven, was the natural bridge by which Iris traveled, so her name was given to the rainbow. The colored portion of the eye, which is differently colored in different people, is also the *iris*, from the word rainbow.

Cobalt

ONLY SEVEN metals were known in ancient times: gold, silver, copper, iron, tin, lead, and mercury. Medieval miners who came across ores of other metals were usually nonplused and at a loss for methods of handling them.

About 1500, for instance, miners in Saxony came across ores which did not smelt properly and which spoiled batches of ordinary ores. The miners had no real understanding of why this should be so, no concept of new metals that required new treatment for isolation. Their explanation was simpler and more direct: earth spirits had bewitched the ore just to be annoying.

One of the German earth spirits is the "kobold." This comes from the primitive Germanic but may have some kinship with the Greek "kobalos," a term used for a mischievous person, from which we ourselves probably get the word *goblin.* In any case, the Saxon miners called the annoying ore kobold.

About 1735, the Swedish mineralogist Georg Brandt, after years of interest in this ore (which had gained some importance in the manufacture of a deeply-colored blue glass), isolated a new metal from it, to which he attached the name given it originally by the exasperated miners, and which has now become *Kobalt* in German and *cobalt* in English and French.

A second ore that also annoyed the long-suffering miners was called by them Kupfernickel. In German, "Kupfer" means "copper," and "nickel" (like "kobold") means a mischievous imp. (We ourselves sometimes refer to the devil as "Old Nick.") Kupfernickel, therefore, means "devil's copper" or "false copper."

In 1751, another Swedish mineralogist, Axel Fredrik Cronstedt, isolated a new metal from this second ore and he, too, kept the miners' name, which was shortened eventually to *nickel.*

Colloid

IN 1861, the Scottish chemist Thomas Graham placed various solutions in a cylinder in which the open-ended bottom was blocked off by a thin sheet of parchment paper. He then placed the cylinder in a bucket of pure water.

If the solution in the cylinder contained any of a number of substances, such as, for example, ordinary salt or sugar, these found their way through the parchment and could be detected in the water in the bucket by some appropriate chemical test.

There were other substances, however, which if present in the solution within the cylinder would never pass through the parchment, no matter how long one waited.

The first set (salt, sugar, etc.) formed thin, watery solutions and happened to exist in crystalline form when not in solution (see CRYSTAL). These, therefore, Graham called *crystalloids*. The second set, which included various proteins, gums, and so on, formed thick gluelike solutions and did not form crystals when not in solution. He called these *colloids*, from the Greek "kolla" (glue).

But later events spoiled the logic of the names, since biochemists have now managed to crystallize many substances that are colloids in solution. The difference between crystalloids and colloids has nothing to do with crystals, actually. It is just that crystalloids have small molecules that can get through the submicroscopic holes in parchment, while the colloids have large molecules (or large aggregates of small molecules) that cannot.

Many of the most important substances in living tissue are colloids in solution, and, to purify them, biochemists still make use of Graham's original experiment. They place a solution inside a membrane (improved over Graham's parchment, of course) and bathe the membranous bag in water. The small molecules escape through the membrane and the large do not so that the two classes of substance are separated. Graham called this process *dialysis*, from the Greek "dialyein" (to separate), and the name remains in use to this day.

Comet

ANCIENT MAN was well aware of the regular movements of the heavenly bodies and knew how they marked off the seasons on earth (see SOLSTICE, EQUINOX) and even thought that the variable but predictable motions of the planets had some influence over human lives. It was a cause for concern and even terror, then, to have something brand-new and of strange appearance suddenly shining in the sky at unpredictable intervals. It must mean the upsetting of the seasons; famine; drought; destruction; catastrophe of one sort or another.

The something that appeared in the sky was not a sharp point like the stars and planets but a fuzzy patch with a long smoky extension. The ancients saw in it a resemblance to a distraught woman fleeing, with her long hair streaming out behind. The Greek word for "long hair" is "kometes." The Romans called the objects "stellae cometae" (hairy stars) and we call them simply *comets*.

The Greek philosopher Aristotle thought the heavens were perfect and could not change. Only the earth and the regions below the moon could show change and corruption. Comets, therefore, must be part of the earth's atmosphere and were not really heavenly objects. (Other ancient philosophers disagreed, but as in other things, Aristotle's ideas, whether right or wrong, usually won out.)

In 1588, however, the Danish astronomer Tycho Brahe proved that the comet of 1577 had been much farther off than the moon. In 1704, the English astronomer Edmund Halley, studying comets, noticed similarities in the paths of the comets of 1531, 1607, and 1681. He declared they were the same comet, and predicted it would return about 1758. (It did, in 1759, seventeen years after Halley's death, and twice since then, in 1835 and 1910.)

This comet, now called *Halley's Comet*, was the first of a number whose orbits have been calculated, so that comets have been reduced to the status of normal members of the solar family and are no portents of disaster at all.

Continent

THE EARLY peoples did not, naturally, have accurate notions of the major features of the earth's surface. If they lived near a shore, they recognized the existence of land and sea, however.

People who had much to do with the sea (the early Mediterranean people, for instance) could not avoid noticing that there were two types of land. There were, first, little pieces of land surrounded by sea. One Latin word for "sea" is "salum" from "sal" meaning "salt." (After all, the remarkable thing about sea water, which is undrinkable, as compared with drinkable river water, lake water and well water is that sea water is salt. We ourselves sometimes call the sea, "the brine.") Anyway, a piece of land in the sea is "in salo" from which comes the Latin "insula" and our *isle*.

Then there was another kind of land, land that went on and on with no sign of sea that the ancients knew of. It was "continuous" land and so was a *continent*, from the Latin "continens" (continuous). In English, we speak of a continent, also, as the *mainland* (*main* coming from the Latin "magnus" meaning "great") as opposed to *island* ("isle-land"). In the days of the Spanish control of the Americas, pirates had their strongholds in the islands of the West Indies and went sailing against the *Spanish Main*.

To the Greeks, there seemed to be three continents separated by sea. The *Mediterranean Sea*, in fact, gets its name from the Latin "medius" (middle) and "terra" (land). It was a sea that lay in the middle with three bodies of land about it. The three continents (Europe, Asia and Africa) are actually connected by land. Africa and Asia are connected by only the small Sinai *isthmus* (from the Greek "isthmos," meaning "a narrow passage") so that may be ignored, but Europe and Asia are connected by land for a thousand miles or more and it is only custom that makes us speak of Europe as a continent. Many geographers speak of Asia plus Europe as Eurasia, while Eurasia plus Africa is sometimes called the World Island. The three together are, after all, surrounded by sea, so they form a large island, and they contain about 85 per cent of the earth's population, so it is an island that is almost the equivalent of the world.

Cortisone

THE ADRENAL glands (see HORMONE) are actually two glands in one, an outer gland enclosing an inner. The inner gland is the *adrenal medulla*, "medulla" being the Latin word for "marrow" or, generally, something which is inside, as marrow is inside bones. The outer gland is the *adrenal cortex*, "cortex" being the Latin word for "bark" or, generally, something which is outside as bark is outside the tree trunk.

The adrenal medulla produces the hormone adrenalin (see HORMONE), while the adrenal cortex produces a whole series of hormones of an entirely different type. Edward C. Kendall, a biochemist at the Mayo Foundation, first isolated these in the 1930's and, before their chemical structure was known, called them simply Compound A, Compound B, and so on.

As it turned out, though, the hormones from the adrenal cortex are all steroids (see CHOLESTEROL) so that they are now called *adrenocortical steroids*. Chemists don't like long names any more than do other people, so this has been shortened to *cortical steroids* and, even further, to *corticoids*.

Some of the corticoids are ketones (see KETONE) so that in naming them, the "-one" suffix, reserved for ketones, can be used. Compound B, for instance, once its structure was worked out, was named *corticosterone* (the "ketone steroid from the cortex," you see).

That name could be used as a starting point. The structure of Compound E was like that of corticosterone, except for a hydroxyl group present at the carbon-17 position and a couple of hydrogen atoms missing at the carbon-11 position. (The carbon atoms of organic molecules are often numbered for convenience according to an agreed-upon system.) Compound E can therefore be called *11-dehydro-17-hydroxycorticosterone*. But this is too long, especially since, in 1948, it was discovered at the Mayo Clinic that Compound E could give amazing relief in some types of arthritis. Anticipating much use of the name, Kendall and his group shortened it by taking certain letters out of the long name and reducing it to just *cortisone*.

Cosmic Rays

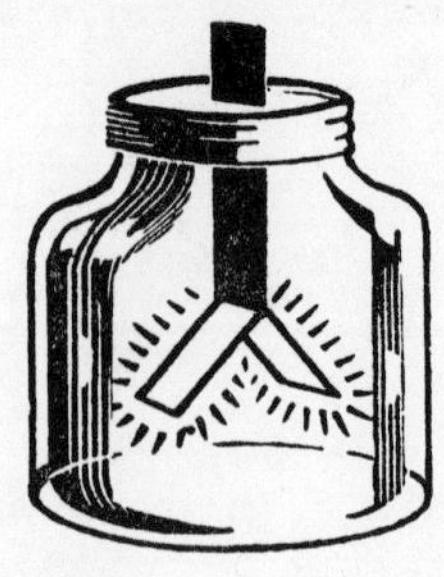

An early way of detecting short-wave radiations was to use a box containing two very light sheets of gold leaf attached to a rod at one end. When the rod is charged with electricity, the two leaves repel one another, taking on a V-shape. In the presence of X rays or gamma rays, electrons are knocked out of the air molecules in the box and the air can then carry electricity. The electric charge leaks off the gold leaf into the air and the two leaves come together. This instrument is an *electroscope.* (The suffix "-scope" comes from the Greek "skopein," meaning "to watch." The instrument uses electricity, you see, to watch for radiation.)

In the absence of X rays, gamma rays, and such, the gold leaves of the electroscope ought to remain apart indefinitely, but they don't. Slowly, they sink together. Scientists assumed there were small quantities of radioactive substances everywhere in the soil (and there are) and that these were the source of perpetual radiation that accounted for the slow leak of the electroscope's charge.

If this were so, however, an electroscope that was miles high, with a thick blanket of air shielding it from the radiating soil, ought to retain its charge indefinitely. Electroscopes were taken up in balloons in 1911 and thereafter, just to prove this and settle the problem so that scientists could forget about it.

But the unexpected happened. Several miles in the air, the electroscope leaked more rapidly than it did on the ground. There was radiation being detected, but not radiation from the ground. The Austrian physicist V. F. Hess was the first to publicize this by way of a name. He called the radiation *Höhenstrahlung*, which is German for "radiation of the heights."

After the First World War, the American physicist Robert Andrews Millikan took the lead in such balloon experiments, and in 1925, he suggested the name *cosmic rays*, since they came not from Earth but from somewhere in the outer *cosmos* (a Greek word meaning "order," particularly "good order," which expresses the Greek idea that the universe forms an ordered whole).

Crucible

THE LATIN word "crucibulum" originally referred to a light that was kept burning before a crucifix (the Latin "crux" means "cross," genitive, "crucis"). When the alchemists melted materials to red-hot liquids in pots, they called the pots by that name because they seemed to hold glowing lights. The word comes down to us as *crucible*. A more entertaining possibility is the story (probably untrue) that alchemists put crosses on their melting pots to keep away devils that might interfere with their experiments. (Any modern chemist will testify that something of the sort is badly needed on occasion.)

Another common vessel used in chemistry is a cylindrical glass container, usually with a lip for pouring. This is called a *beaker*, which might seem to be derived from the fact that the lip looks like a bird's beak, but which actually comes from the Greek "bikos," meaning a "jar" or "cask." Similarly, a glass container which narrows at the top to a small opening is called a *flask*, from the Latin "flasca" meaning a "wine bottle."

Oddly enough, the most familiar chemical vessel to the average nonchemist is one that went out of fashion a century or more ago. This vessel is a flask with a long, narrowing neck that is bent over to one side, slanting downward. It has the purpose of collecting vapors. That is, a liquid heated in the flask will give off vapors which will travel down the neck away from the flame and cool to liquid. The drops of liquid formed will drip out the end of the neck and be collected. Such an instrument is called a *retort* from the Latin "re-" (back) and "torquere" (to twist) because the neck of the vessel is "twisted back" or, in Latin, "retortus."

The vessel is inefficient because the neck gradually warms up so that after a while the vapors no longer turn to liquid. Retorts have been replaced by devices in which the tube through which the vapor passes is water-cooled. The very word *retort* is no longer used in the chemical laboratory for any vessel. Nevertheless, the alchemist's retort has an unshakable association with chemistry in the popular mind and forms the major part of the insigne of the Chemical Corps of the United States Army.

Cryptogamous

THE TYPICAL and most advanced land plants are the spermatophytes (see ANGIOSPERM) but there are common land plants that are more primitive. Of these, the best known are the ferns, which, like the higher plants, have leaves, stems, and roots. They do not, however, have flowers or seeds.

Ferns reproduce themselves, instead, by means of spores, which are like seeds in that they will develop into adult plants, but unlike seeds in that they lack within them the tiny leaves and other differentiated parts that the true seeds of the higher plants have. (Nevertheless, *spore* is another form of the word "sperma" meaning "seed" (see ANGIOSPERM), both "spore" and "sperm" come from the Greek "speirein," "to sow.")

The leaves of ferns are quite different from those of higher plants. In ferns, individual leaflike extensions called *fronds* (from a Latin word, "frons," meaning "leaf") come off a central stalk so that the whole leaf rather resembles a feather. In fact, the Greek word "pteris," meaning "fern," comes from their word "pteron," meaning "feather." And the plant group to which ferns belong are today called *Pteridophyta*. Since the Greek "phyton" means "plant," ferns are "feather-plants." The word *fern* itself has been traced back to the Sanskrit "parna" which means "feather," among other things.

A still more primitive group of land plants includes as their most familiar representatives the mosses. The Greek word for "moss" is "bryon" so this group is called *Bryophyta* ("moss-plants").

Like the ferns, the mosses lack flowers and true seeds and reproduce by means of spores. For this reason, an older classification, introduced by Carl von Linnaeus in the 1700's, listed both groups under the heading of *Cryptogamia* from the Greek "kryptos" (hidden) and "gamos" (marriage). It is by means of flowers, you see, that higher plants "marry" (i.e., are pollinated) and produce seeds. To produce the equivalent of seeds (i.e., spores) without flowers was to have a "hidden marriage" — no visible evidence. This is now an outmoded group name but ferns and mosses are still spoken of as *cryptogamous* plants.

Crystal

The atoms in a solid may have no particular arrangement and the solid is then said to be amorphous (see DIAMOND). More often, however, the atoms are arranged in a definite pattern, and this results in the solid itself taking on a certain *symmetry*. (The Greek prefix "sym-" or "syn-" means "with" and implies things behaving in cooperation and not in conflict. They are "with" and not "against." The word "metron" means "measure"; thus symmetry implies that the various measurements of an object combine smoothly and do not clash.)

The Greeks saw the best example of this in the case of snowflakes or the formation of hoarfrost patterns. The Greek word for "frost" was "kryos," so they called these patterns of snow and ice "krystallos."

Another remarkable thing about ice was that it could be transparent and since the Greeks knew very few other transparent objects, that property struck them forcibly. Consequently, when they found pieces of rock that had symmetrical shapes and were transparent they called them "krystallos," too, and considered them a form of ice.

By early modern times, however, it was realized that many solids could take on symmetrical shapes if allowed to solidify slowly out of a solution, or out of molten form. They were not forms of ice and they did not have to be transparent. It was the symmetry that counted. The word (*crystal*, in English) was applied to all of them.

It is also applied to glass objects which are cut into symmetrical shapes, even though glass itself is amorphous and its symmetry is purely artificial and not a natural reflection of atom arrangement.

In fact, sometimes it is the transparency of an object and nothing more that forces the use of the word. We speak, for instance, of a fortuneteller's "crystal" ball. This is nothing more than a glass sphere with no crystallinity about it at all, either real or artificial. It is "crystal" only because it is transparent and because the Greeks were once amazed at the phenomenon of transparency.

Cumulus

The visible sign of water vapor in the air is the *cloud*. This is derived from the Anglo-Saxon "clud," meaning any round mass as, for instance, a stone. The word *clod* for a lump of hard earth comes from the same root. To primitive people, unaware of the nature of clouds, water vapor or air, it isn't so strange that a cloud might well look like a pebble in the sky.

An ordinary cloud is a suspension of tiny liquid droplets in air; it might also consist of tiny solid particles, in which case it is *smoke*. This comes from the Anglo-Saxon, but there is a Greek word, "smychein" (to smolder). The word *smother* is also related to both *smoke* and *smolder*.

A cloud which happens to be at the earth's surface is *fog* or *mist*. The former is a Danish word meaning "spray" or "driving snow," the latter an Anglo-Saxon word meaning "darkness." These are picturesque derivations. Fog and mist do indeed obscure the sun and bring on a premature darkness, just as driving water (spray) or snow would. In some industrial areas, such as Los Angeles, smoke may mix with a persistent fog, and a combination word, *smog* (*sm*oke-*fog*) has been invented to describe this, and popularized by Hollywood comedians.

The most beautiful cloud is the "woolpack," the typical cloud of fair summer weather. It results from a column of warm air rising and meeting colder air aloft, so that water vapor condenses into fine droplets that look like a mass of wool heaped up high. These are *cumulus* clouds, "cumulus" meaning "heap" in Latin.

If the air contains enough moisture, then, as it rises (or comes in from the sea), it will form a cloud that will spread over the visible heavens and be thick enough to let so little sunlight through as to take on a gray and threatening appearance. These are *nimbus* clouds, "nimbus" being Latin for "rain clouds" and, of course, such clouds often result in rain.

Sometimes very high clouds form in the shape of closely spaced feathers or tufts of hair. These are *cirrus* clouds, "cirrus" meaning "a curl of hair."

Cyanide

COLORED INORGANIC compounds suitable for use in various types of ornamentation have always been valuable. One blue mineral much used for decoration was *lapis lazuli.* "Lapis" is Latin for "stone" while "lazuli" is apparently a corruption of the Arabic "lazaward," meaning "sky blue" (*azure* is another corruption); lapis lazuli is "azure stone." Powdered lapis lazuli was called *ultramarine*, from the Latin "ultra" (beyond) and "marinus" (the sea) because it was imported from "beyond the sea."

When, in 1704, two Berlin dyers accidently discovered a new deep-blue compound of iron, a kind of substitute lapis lazuli, they kept its method of preparation secret so that it could be called only *Prussian blue* from its color and place of origin.

Naturally, the secret could not be kept forever. Chemists got to work. In 1783, the Swedish chemist Karl Wilhelm Scheele obtained a weak acid from Prussian blue which he called *prussic acid.* In 1815, the French chemist Joseph Louis Gay-Lussac isolated a gas from another source which could be easily converted to prussic acid. He found that the key atom group in these compounds consists of a carbon atom and a nitrogen atom. Prussian blue contains six such combinations so he called this key group the *cyanide group* from the Greek "kyanos" (blue). In prussic acid, the cyanide group was attached to a hydrogen atom so that became *hydrogen cyanide.* The gas he had discovered contained two cyanide groups in its molecule and this he called *cyanogen.* (The "-gen" suffix is generally used by chemists to mean "give birth to" or "produce." See HYDROGEN. Cyanogen is a substance that "produces cyanide.")

Because compounds like cyanogen, hydrogen cyanide, and potassium cyanide are deadly poisons, the "cyan-" stem has come to have an unpleasant sound and yet it is used in many innocent words for the sake of its "blue" significance. There is a harmless blue compound called *cyanidin*, and a *cyanometer* (the Greek "metron" means "measure") does not measure poisonousness by any means, but only the intensity of the blue of the sky.

Cyclone

WHAT WE call weather is largely a matter of air movements. Cold air blows down from the north and warm air from the south. Moist air blows in from the ocean, forming clouds and bringing rain. Dry air blows in from the interior and brings drought. *Weather* is an Anglo-Saxon word that has been traced back to an old word that may be related to the Slavic "vetra," meaning "wind." This is a reasonable theory. English folk-idioms frequently make use of synonyms as follows: "spick and span," "nook and cranny," "hale and hearty" (often using old words that haven't survived in any other fashion). Well, we speak of "wind and weather."

Because of the importance of wind, there are a number of words for it in all its forms, most (such as *breeze*, *gust*, *blast*, *storm*, etc.) being of Anglo-Saxon derivation, stretching back to old Germanic words. There are exceptions, though. A gentle wind is a *zephyr*, from the Greek "zephyros" (west wind), since to the Greeks the west wind brought neither snow from the north nor heat from the Saharan south, and was thus a gentle wind.

In some ways, weather depends upon seasons. There are hot, cold, wet, and dry seasons. There are times of the year when it is particularly apt to be stormy and the word *tempest* attests to that since it comes from the Latin "tempus" (time). In Latin, "tempestas" means both "period of the year" and "storm."

The severest forms that wind can take involve circular motions. These begin when cold air and warm air meet and get to whirling because of the motion of the earth. (A point near the equator moves more quickly than a point farther removed from the equator.) In the northern hemisphere, then, the southern part of two colliding masses is dragged east more quickly than the northern part and a *counterclockwise* whirl — opposite in direction to the motions of the hands of a clock, from the Latin "contra" meaning "against" — is set up; a circular wind, called a *cyclone*, from the Greek "kyklos" (circle). In the southern hemisphere, the northern portion moves faster, and a clockwise wind is set up, an *anticyclone* from the Latin "anti-" meaning "against" or "opposite to."

Cyclotron

IN ORDER to change one kind of atom into another, subatomic particles must be sped into an atom's nucleus with great force. When such atomic changes were first brought about in 1919, the natural particles emitted by radioactive elements (see ALPHA RAYS) were used.

These were insufficient, however, and instruments were devised to speed up subatomic particles in greater quantities to greater energies. These instruments were popularly called *atom-smashers.*

A particularly successful type of atom-smasher was invented by the American physicist Ernest O. Lawrence. He rigged up a device, in 1930, which forced protons into a circular path between two magnets. The protons were driven into greater and greater speed by the magnetic field so that they spiraled outward until, finally, they skidded out of the instrument altogether at a terrific velocity.

The instrument was named a *cyclotron.* The prefix "cyclo-" is frequently used in scientific names, coming from the Greek word "kyklos," meaning "circle." In this case, it referred to the circular path of the protons. The "-tron" suffix is a false analogy with the names of some of the subatomic particles — electron, neutron, and so on.

The "-tron" suffix has become customary in naming new atom-smashers that have been developed since the cyclotron. For instance, in 1940, the American physicist D. W. Kerst designed an instrument to accelerate electrons to great speeds. Since a speeding electron is a beta particle (see ALPHA RAYS), the new instrument was named a *betatron.*

The energies of a speeding particle are measured in electron volts, which is abbreviated "ev." At the University of California, an atom-smasher has been built that will speed up particles till they have energies of billions of electron volts. A billion electron volts is abbreviated "bev" so the instrument is called a *Bevatron.* Since such energies are about those of cosmic rays, a similar instrument at Brookhaven is called a *Cosmotron.*

Cystine

About twenty different amino acids have been found in proteins (see glycine), their discoveries being spread over a century and a quarter. Yet the first to be discovered wasn't connected with proteins for nearly ninety years.

It happened that in 1810, an English physician and chemist William Hyde Wollaston was analyzing a bladder stone that had been removed from a human patient. (These "stones" form occasionally in kidney and bladder from insoluble substances that precipitate out of urine. There are different varieties of these stones and Wollaston, as it happened, had gotten hold of one of the rare types.) He found the stone to be made up mostly of a sulfur-containing organic compound which he named *cystine*, from the Greek "kystis" (bladder). It was not till 1899 that the same amino acid was located in horn. Horn contains a protein, *keratin* (from the Greek "keras," meaning "horn") which is, of all proteins, the richest in cystine, so that was only right.

Then a similar amino acid was discovered into which cystine could be easily converted. The second amino acid was named *cysteine* to emphasize the similarity, but the additional *e* makes little impression on the eye and none on the ear, and that is entirely too similar for comfort.

Others of the protein amino-acids were also named after the object in which they were first found. For instance, one was isolated from cheese in 1849 and named *tyrosine* from the Greek "tyros" (cheese). Then, another was isolated from silk in 1865 and named *serine* from the Latin "sericus" (silken), which in turn comes from "Seres," the name of a people in Eastern Asia.

More indirectly, an amino acid was isolated from asparagus in 1806 and named *asparagine*. In 1832, asparagine was converted into a closely related compound that was a stronger acid and was therefore named *aspartic acid*. It was not until 1875, however, that aspartic acid was recognized as one of the amino acids in proteins, and later still that asparagine also was added to the list.

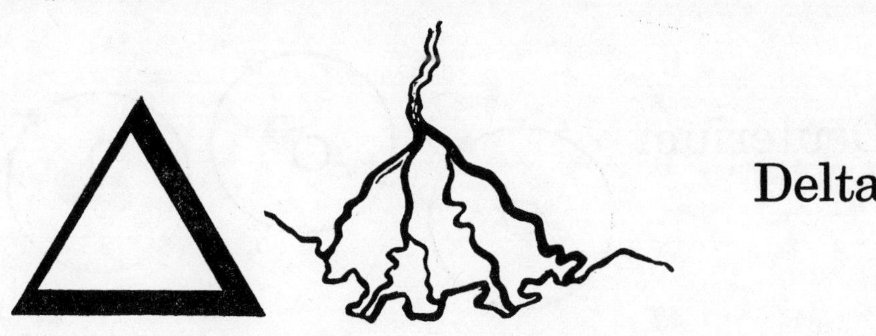

Delta

THE LARGE RIVER best known to the Greeks of the time of Herodotus was the Nile of Egypt. Herodotus spoke of it with admiration and called Egypt "the gift of the Nile." The reason for that was, first, that the Nile was an unfailing flow of water through a rainless desert and, second, that once a year it overflowed its banks, leaving behind, as it receded, new and fertile soil. This soil is carried down from the highlands of East Central Africa where the sources of the Nile are located.

The turbulent floodwaters deposit loosened soil and silt along its lower course but manage to carry some right down to the Mediterranean. The slower the current, the less material can be carried by a river and at the sea, where the current flow stops altogether, all that remains of the silt is dropped.

There are no tides in the southern Mediterranean to carry the silt away, so year after year it collects at the mouth of the Nile, and the Nile must find its way over the silt to the slightly more distant Mediterranean where it again deposits silt. In this way, a whole area of flat fertile soil is built out into the sea, across which the river splits up into slow-moving branches.

To Herodotus, looking across the sea to Egypt, the mouth of the Nile looked like the illustration above.

Now it is customary for us to name things after letters if they have the proper shape (U-curve, I-beam, T-square and so on). The Greeks did the same. The triangular area resembled the fourth letter of the Greek alphabet (Δ) and so, Herodotus tells us, they called it by the name of the letter, *delta.* The word is now used for all such built-up mouths, even when there is nothing triangular about them. The Mississippi delta, for instance, is shaped nothing at all like the Greek delta, as you will see if you look at the map.

Deuterium

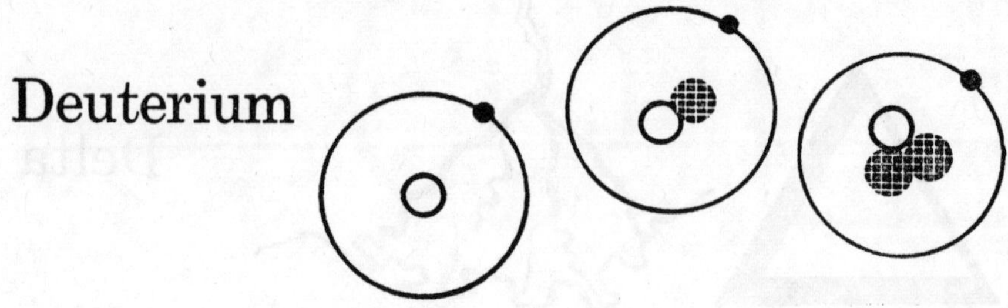

Isotopes are varieties of atoms of an element that differ among themselves in their weight. For instance, some oxygen atoms weigh 16 units, some 17, and some 18. They are distinguished by being called oxygen-16, oxygen-17, and oxygen-18.

In 1931, however, the American chemist Harold C. Urey, and co-workers, discovered an unusual isotope. It was a hydrogen isotope present in small quantities wherever ordinary hydrogen occurred. Ordinary hydrogen, the lightest of the atoms, has a single proton (unit weight) in its nucleus. It is *hydrogen-1*. The new isotope had a proton (unit weight) plus a neutron (also unit weight) in its nucleus and was *hydrogen-2* or, more colloquially, *heavy hydrogen*. An atom of hydrogen-2 weighed twice as much as one of hydrogen-1. There was the unusual property. No other isotopes in the list of elements differed so in weight (on a per cent basis).

Because of this large weight-difference, hydrogen-1 and hydrogen-2 are, for isotopes, unusually distinct physically and chemically. They seemed to deserve special names. The British physicist Ernest Rutherford suggested *haplogen* for hydrogen-1 and *diplogen* for hydrogen-2 from the Greek words "haploos" (single) and "diploos" (double).

Urey, however, suggested *deuterium* for hydrogen-2 from the Greek word "deuteros" (second) and it was this name that was adopted. By analogy, hydrogen-1 is *protium* from the Greek word "protos" (first). Then, when the British physicist M. L. E. Oliphant discovered a still heavier isotope, *hydrogen-3*, in 1934, it was automatically named *tritium*, from the Greek word "tritos" (third).

Since the nucleus of a protium atom is a proton, the nucleus of a deuterium atom (a one-proton-one-neutron combination) is called a *deuteron*, while the nucleus of a tritium atom (a one-proton-two-neutron combination) is called a *triton*.

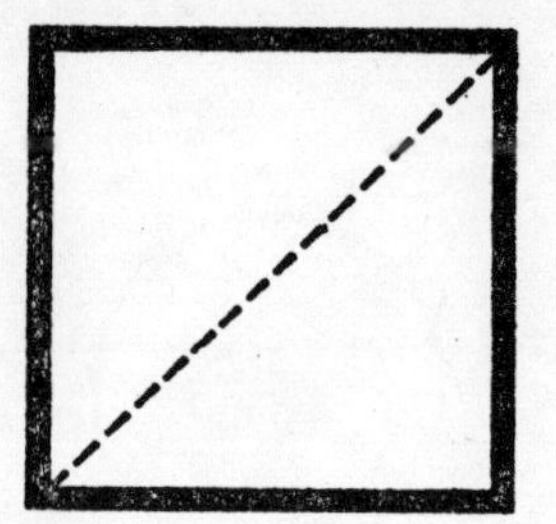

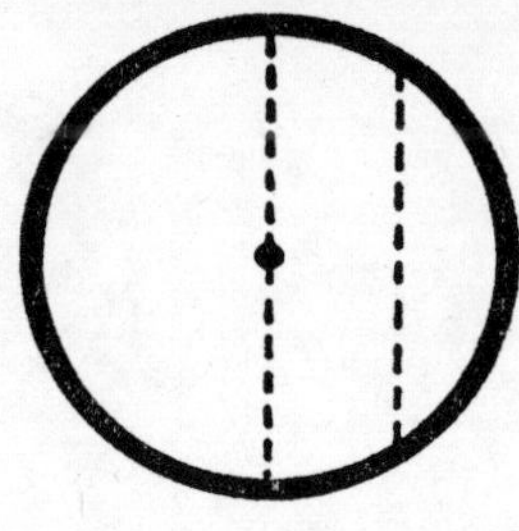

Diagonal

In any polygon, two adjacent angles are connected by sides of the figure. If the polygon has more than three sides, then it is possible to connect two angles that are not adjacent by a straight line that cuts directly across the polygon. Such a line is a *diagonal*, from the Greek prefix "dia-" (through) and "gonia" (angle); it passes "through the angles."

The most familiar polygon is, of course, the square, and this is usually drawn so that two sides are horizontal and two vertical. In a square drawn this way, the diagonal, moving from one angle to the opposite, must run slantwise. *Diagonal* has therefore come to mean a slantwise direction, though the proper word for that is really *oblique*, which comes from the Latin "obliquus" — "ob-" (before) and "liquis" (crooked). An oblique line, in other words, runs crookedly on before, slanting away from true.

A circle has no angles and can have no diagonal, and yet a straight line can be drawn across the circle from one side to the other. Such a line is called a *chord* from the Greek "chorde," (the innards of an animal). The intestines were used, of course, to make the strings of musical instruments, which points up another meaning of the word (see octave), but strings of some sort were the first measuring instruments, and a lute's strings would do in a pinch.

A chord that passes through the circle's center is the longest possible chord for a given circle. It measures the greatest thickness of the circle. This is the *diameter*, from the Greek "dia-" (through) and "metron" (a measure); it is a "measure through" the circle.

The sum of the lengths of the sides of a polygon is, similarly, the *perimeter*, from the Greek "peri-" (around) and "metron" (a measure); it is a "measure around" the polygon. This could, without illogic, be used for the length of the boundary of a circle, too, but in the latter case, mathematicians, for some reason, use the Latin equivalent. The length of the circle's boundary is the *circumference*, from the Latin "circum-" (around) and "ferre" (to carry). The measure is "carried around" the circle.

Diamond

Occasionally, an element can exist in two or more different forms, depending on temperature, pressure, or other environmental factors. As an example, we have oxygen and ozone (see bromine). Generally, one such form can be changed into another in the laboratory and the Swedish chemist J. J. Berzelius, in 1841, suggested these forms be termed *allotropes* of an element, from the Greek "allos" (other) and "trope" (change).

The most dramatic example involves the element *carbon*, which is most familiar to us as coal, some forms of which are almost pure carbon. The Latin word for coal was "carbo" (genitive, "carbonis"), hence the name of the element. The word *coal* was originally applied to any burning ember. If wood is heated without being allowed to burn, for instance, a black residue is left, which will burn slowly. This is *charcoal*, i.e., coal formed by charring.

Coal, charcoal, and various forms of soot are allotropes of carbon in which the carbon atoms are arranged in no particular order. They form no regular pattern or shape. A substance such as coal is therefore said to be *amorphous* (from the Greek "a-" meaning "no" or "not" and "morphe" meaning "shape").

Carbon atoms can arrange themselves into rings of six, joined in sheets that resemble an orderly array of bathroom tiles, each sheet held loosely to neighboring sheets. Such sheets are easily split apart and if some of this material is rubbed over paper, pieces flake off and remain behind. This is the secret of the pencil, and this allotrope of carbon is called *graphite*, from the Greek "graphein" (to write).

If carbon is put under great temperature and pressure, the atoms take up a very symmetrical arrangement that holds together unusually tightly. The result is a transparent and extremely hard allotrope of carbon called *diamond*. This word is a corruption of the word *adamant*, once used for the jewel. *Adamant* in turn comes from the Greek "a-" (not) and "daman" (to subdue). The jewel was so hard, you see, that it could not be subdued. Nothing else, that is, could scratch or make an impression upon it.

Digitalis

As EVERY child knows, fingers are handy when it comes to solving simple mathematical problems, such as adding two and four. This shows up in the words we use. The Latin word for a finger or toe is "digitus" and we still call fingers and toes digits. But the word *digit* also refers to whole numbers — one, two, three and so on. Fingers and numbers are so closely related that the same word will do for both.

Digits come into medicine, too. There is a European plant, with drooping purple or white flowers, that in English is called foxglove. In German, the name is *Fingerhut*, which means "thimble" because the flowers form little tubes that look like thimbles.

In 1541, a German botanist, Leonard Fuchs, decided to give the plant a Latin name (only Latin names are official for plants and animals). With *Fingerhut* in mind, he called the group of plants to which the foxglove belongs *digitalis*. In Latin, this means "of or pertaining to the finger," and certainly a thimble is something that pertains to a finger.

Now, in the days before modern medicine, there were people who cooked up various plants according to secret recipes handed down from generation to generation and used the results to treat sickness. One of the plants valued for this purpose was the foxglove. However, as medicine advanced, doctors laughed at this sort of superstitious nonsense.

Just the same, an English physician, William Withering, wrote an article about the medicinal uses of foxglove in 1785 and was not ashamed to admit that he grew interested in the plant as a result of information received from an old country woman, whose family secret it was.

And in fact for nearly two centuries now, chemicals obtained from the foxglove have been used to improve the wavering heartbeat, slowing it and making it more even and intense. And from the finger, by way of the thimble, the medicines are the *digitalis glycosides*.

Dimension

GEOMETRY BEGAN as the practical art of measuring land areas (the word comes from the Greek "ge," meaning "earth" and "metron," meaning "measure") and determining the volumes of containers. It was important to know how many measurements were needed to fix an area or a volume. This thought is expressed in the word *dimension*, which comes from the Latin "dimensio" — "di-" (apart) and "metiori" (to measure).

A rectangle, for instance, must be taken apart, so to speak, into two measures — its length and its width. The two measures multiplied give the area. A rectangle, therefore, is a two-dimensional figure.

The basic two dimensional figure is the plane, which can be visualized as a continuous sheet without thickness, spreading out in all directions, perfectly level and smooth. (*Plane* comes from the Latin "planus," meaning "flat" or "level.") Any figure which can be drawn on a plane is two-dimensional.

The true mathematical plane does not exist in the real world because nothing real can have zero thickness. An object which has thickness as well as length and width can exist. To measure the volume of a cube, for instance, one must measure all three: length, width and thickness. A cube is a three-dimensional figure.

The importance of reality is shown by the fact that any three-dimensional object is called a *solid*, from the Latin "solidus" (dense). A solid, you see, is dense; it has weight and substance; it is not a mere abstraction.

If we add a fourth dimension, we are away from commonplace reality again. Mathematicians can suppose figures with any number of dimensions and even find it useful to do so. In Einstein's theories, for instance, time can be considered a fourth dimension, though it is not experienced by our senses in the same way as are the ordinary three dimensions of space. To talk of a universe which contains such a fourth dimension, scientists use the expression *space-time*.

Dinosaur

OF ALL THE extinct forms of life on earth, the most dramatic are the gigantic reptiles that lived in the Mesozoic era (see FOSSIL). The most dramatic of modern-day reptiles are the snakes, whose characteristic method of movement gives the name to the whole group of animals, since *reptile* comes from the Latin "repere" meaning "to creep." The word *serpent*, incidentally, comes from the Latin "serpere," also meaning "to creep." The reptiles include all cold-blooded animals with scales and bones and, while some do creep, others swim, leap, and even fly.

The extinct giant reptiles are usually lumped together in the popular mind as *dinosaurs* (though, from a scientific standpoint, it should be realized that some dinosaurs were as small as hens, while some giant reptiles were not dinosaurs), from the Greek words "deinos" (terrible) and "sauros" (lizard) — and "terrible lizards" they were, indeed.

The most terrible dinosaur was *Tyrannosaurus Rex*, which grew to as much as 45 feet in length and stood as high as a giraffe on its tremendous hind legs, with a skull four feet long and a cavernous jaw equipped with foot-long teeth. It was the largest meat-eater ever to inhabit dry land, and its name is very descriptive.

To the Greeks, it is believed, a "tyrannos" was any individual who gained one-man rule of a city by rising from the common people (like a modern dictator) rather than by inheriting the rule as a member of a kingly line stretching back into antiquity. A tyrannos was not necessarily a worse ruler than a king, but in 510 B.C. the Athenians expelled the tyrant Hippias and established a democracy. After that, they spoke ill of tyrants in general (as we spoke of kings after we defeated George III) and *tyrant* took on its modern meaning. Actually, though, "master" would be a better translation of the original "tyrannos."

Since "rex" is the Latin word for "king," Tyrannosaurus Rex means "King Master-Lizard," which is an excellent name, for if ever there was a king of beasts, this was it.

Dirigible

In 1782, two French brothers, Joseph Michel Montgolfier and Jacques Etienne Montgolfier, lit a fire under a large, light bag with an opening underneath, and allowed the hot air to fill it. The hot air, being lighter than cool air, lifted the bag and this was the first *balloon*, a word which arises from the same source as *ball*, the "oon" ending implying largeness. A balloon is just a kind of large ball after all.

It was not until 1852, however, that another French inventor, Henri Giffard, was able to mount a steam engine in the gondola under a cigar-shaped balloon, and exert enough power in this way to guide the balloon against the wind. This was a *dirigible balloon*, from the Latin "dirigere" (to guide). The phrase was shortened simply to *dirigible*, which was then applied to powered balloons in general.

The man most responsible for making such dirigibles practical was Count Ferdinand von Zeppelin, a German inventor, who first built a large aluminum framework within which a bag might be expanded, thus making giant dirigibles possible. In his honor, such dirigibles are often called *zeppelins*. In fact, the most successful dirigible ever flown was the *Graf Zeppelin*, which made numerous crossings of the Atlantic and circumnavigated the world. It was named in honor of the Count, since *graf* is German for "count."

Still another common name for the dirigible is *airship*, since it floats and moves through air, as an ordinary ship floats and moves through water.

In 1903, for the first time, a machine was driven through the air without a bag of hydrogen to hold it up. It was heavier than air and that was the remarkable thing about it. What did hold it up was the moving air lifting upward under the large *plane* areas of the wings (*plane* being derived from the Latin "planus," meaning "flat" or "level"). The device was therefore named an *airplane* or, using the Greek "aer" (air), *aeroplane*. (A *seaplane*, however, is not one that moves through the sea, but is an airplane equipped to land on the sea.)

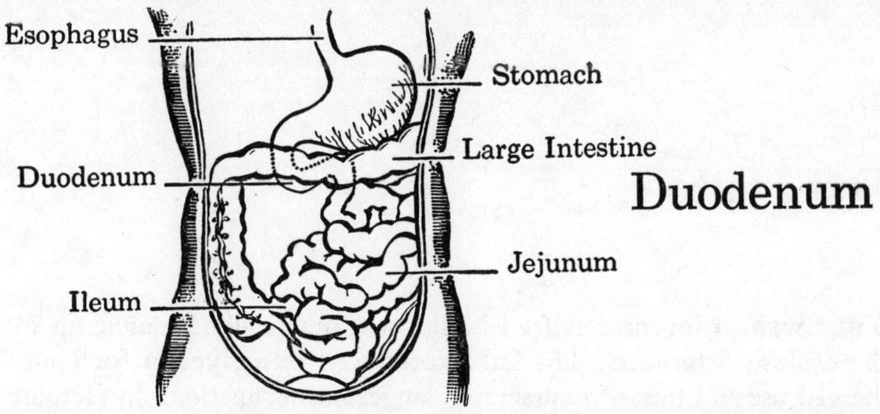

Duodenum

As soon as you put food or drink into your mouth, it has entered the *alimentary canal* (from the Latin word "alimentum," meaning "nourishment"). Once you swallow, the food enters a ten-inch tube that leads to the stomach. The tube is the *gullet*, from the Latin "gula" (throat), or *esophagus*, a Greek word coming, possibly, from "oisein" (to carry) and "phagein" (to eat). It is a tube, in other words, that carries what you eat.

Leaving the stomach, the food enters a long coiled tube called the *intestines*, from the Latin "intestinus" (inner). The last part of the intestine is wider in diameter than the first part, so that we have a long *small intestine* and a comparatively short *large intestine*. (The terms *small* and *large* refer to diameter rather than to length.)

Anatomists have divided the small intestine into three divisions. The first eleven inches of its length is the *duodenum*. This comes from the Latin "duodeni" (twelve) because the early anatomists measured by using their hands as convenient rules and the duodenum is twelve finger-widths in length. The Germans, who are much less patient with classical derivations than we are, call the duodenum *zwölffingerdarm*, which is straight German for "twelve-finger-intestine."

The next eight feet of intestine is the *jejunum*, from the Latin "jejunus" (empty), because the Roman medical writer Celsus thought it retained no food but merely sent it on. The remainder of the small intestine is the *ileum*, a Latin word that comes perhaps from the Greek "eilein" (to roll up) because that is what must be done to the small intestine if it is to be made to fit in the abdomen.

The intestines are also referred to as the *bowels*. This comes from an Old French word "boel" which is, in turn, derived from the Latin "botellus" (a little sausage) and the intestines do look like a string of sausages, at that.

Dynamite

THE SWEDISH inventor Alfred Nobel was practically brought up in the explosives business. His father produced nitroglycerin for commercial use and this, of course, is a dangerous occupation. In fact, an accidental explosion of nitroglycerin killed Nobel's younger brother.

Nobel therefore began to seek ways in which to make nitroglycerin safer to handle. In 1862, he found the answer in a kind of earth called *kieselguhr*. This name is German from "kiesel" meaning "flint" (a very common rock) and "guhr" meaning "an earthly deposit."

Kieselguhr is made up of microscopic pieces of silicon dioxide (of which flint is one form) that once made up the skeletons of tiny one-celled plants called *diatoms*. These plants are so called because, in some common varieties, each cell is divided into two nearly separate parts. The Greek "dia-" means "through" and "temnein" means "to cut"; the cell is almost "cut through," in other words. In English, kieselguhr is called *diatomaceous earth*, or *diatomite*.

Because kieselguhr is composed of individual skeletons with microscopic spaces within and between the pieces, it is porous and will absorb liquids. For instance, it will absorb as much as three times its own weight of nitroglycerin. The mixture of kieselguhr and nitroglycerin can be molded into sticks which can then be handled with practically no danger of accidental explosion. When the sticks are detonated, however, in the appropriate manner, as for instance by an electric spark (set off from a distance), a powerful explosion results. This safe nitroglycerin was patented by Nobel in 1862 under the name *dynamite* from the Greek "dynamis" (power).

Nobel made a large fortune from dynamite and other inventions in the explosives field and when he died he left over nine million dollars for the establishment of the Nobel Foundation. This awards each year the *Nobel Prizes* in five classifications: physics, chemistry, medicine and physiology, literature, and peace. In 1957, element 102 was discovered at the Nobel Institute and named *nobelium* in its honor and that of Alfred Nobel.

Dynamo

THE FIRST method for producing an electric current involved batteries (see VOLT), but batteries are not practical as sources of current on a large and continuous scale. The key to something better came when the Danish physicist Hans Christian Oersted discovered, in 1820, that a wire through which a current was flowing could attract a compass needle, and that therefore electricity and magnetism were somehow related.

The English chemist Michael Faraday showed the reverse to be also true. In 1831, he discovered that a magnet moving through a coil of wire set up a current in the wire. Moving electricity resulted in magnetism and moving magnetism resulted in electricity.

It was then only necessary to think up a way of rotating a coil of wire between the poles of a magnet (it doesn't matter whether magnet moves and coil is stationary or vice versa) and then bleeding off the electricity as it was formed. The coil can be made to turn continuously by means of a turbine run by water or steam (see ENGINE). In this way, mechanical energy is converted into electrical energy in sufficient quantities to light cities and run huge factories.

Such a device is called a *generator*, from the Latin "gignere" (to produce); it certainly "produces" electricity. An earlier name, however, was "dynamo-electric machine." The Greek "dynamis" means "power," so the longer name means "a machine producing electricity from ordinary nonelectric power," so to speak. That name was shortened to simply *dynamo*.

This name had an unfortunate consequence for the city of Constantinople. When it was first suggested that that city be electrified, it was explained to the Sultan of Turkey that it would be necessary to install dynamos. The Sultan, who was not a man of advanced education, knew only that *dynamo* sounded distressingly like *dynamite* and he knew what dynamite was. So he vetoed the project and Constantinople had to wait several additional years for its electricity.

Echinodermata

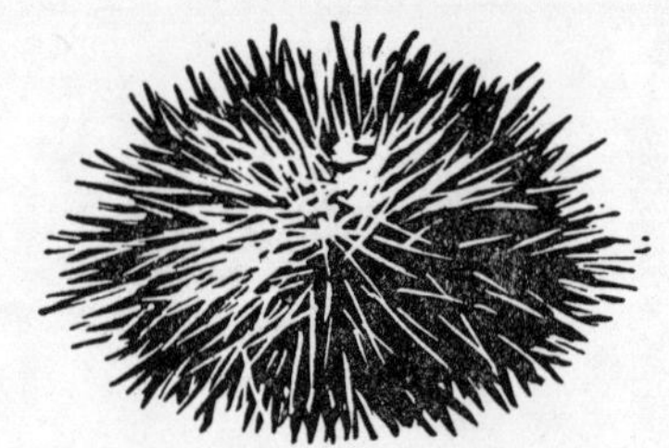

It was an old medieval theory that for every animal on land, there was an equivalent animal in the sea. Some of that theory remains in the names given various sea animals. For instance, certain species of seals are called sea lions, sea bears and sea leopards, while a particularly large seal with a bulging snout is the sea elephant. The manatee, a sea mammal not related to the seals, is usually called the sea cow, while a porpoise (a small member of the whale family) is sometimes called a sea hog.

The fish also contribute examples, the most famous being the sea horse, a small fish with a head that resembles a surrealist version of a horse. There is also a fish with a spiny skin, capable of blowing itself up into a sphere when threatened. It is then both difficult and dangerous to grab. It is called the *globefish*, in consequence, or the *sea hedgehog*.

The term *sea hedgehog* also applies to another creature, a much more primitive one. It is an invertebrate animal (see PHYLUM) in the shape of a sphere, flattened on one side. It rests on the sea bottom on that flat side (in the center of which is the mouth). The rest of the body is thickly covered with spines. The Latin word for "hedgehog" is "ericius," from which we get *urchin* (usually applied, these days, to little boys, who can indeed have a prickly nature and be hard to get close to without damage) so the spiny sea creature is called a *sea urchin*. The Greek "echinos" means "hedgehog," and the class of organisms to which the sea urchin belongs is *Echinoidea*.

The sea urchins, along with other creatures with spiny skins and additional features in common, belong to the phylum *Echinodermata*, from the Greek "derma," genitive, "dermatos" (skin), hence the "spiny-skins." The best-known examples of the phylum are the various *starfish* (these are shaped like stars, but are not fish) which belong to the class *Asteroidea*, from the Greek "aster" (star). Plants as well as animals may have their equivalent in the sea, I should mention, since there are members of this phylum which are commonly called, owing to their appearance, *sea cucumbers* and *sea lilies*.

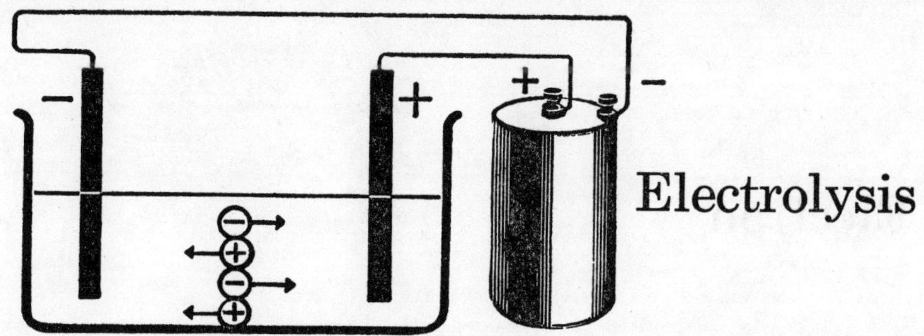

Electrolysis

ONCE VOLTA invented the battery (see VOLT), chemists had electric currents to play with. If such a current were passed through certain liquids, chemical changes took place. Generally, there was a separation of the molecules of the substances in solution into smaller pieces. From dissolved copper sulfate, copper could be plated out. Hydrochloric acid in solution would give off chlorine and hydrogen; water would split up into oxygen and hydrogen, and so on.

Because the molecules seemed to loosen and break apart, the process was called electrolysis. Since the Greek word "lysis" means "a loosing," *electrolysis* is "a loosing by electricity."

Pure water will not carry an electric current so any substance such as sulfuric acid or sodium chloride which, when added to water, allows an electric current to pass and electrolysis to proceed is called an *electrolyte*. A substance such as sugar, which, when dissolved, does not allow the electric current to pass is a *nonelectrolyte*.

In order for an electric current to pass through a liquid, two metal rods — one connected to the positive pole of a battery, and one to the negative pole — must be dipped into the liquid. These two rods are *electrodes*, the "-ode" suffix coming from the Greek "hodos," meaning "route." The electrodes form the route of the electric current.

The electrode connected to the positive pole of the battery is the positive electrode; the other, the negative electrode. The British physicist Michael Faraday first suggested (in 1834) that the positive electrode be called the *anode*, the negative, the *cathode*, from the Greek prefixes "ana-," meaning "up," and "kata-," meaning "down." At that time, it was believed the electric current traveled from the battery's positive pole to its negative pole, just as a water current travels from a hilltop to a valley; hence the choice of prefixes. (As a matter of fact, just the reverse is now known to be true; electricity, or at least electrons (see ELECTRON), travels from negative pole to positive.)

Electron

THE ANCIENT Greeks noticed, back in 600 B.C. or thereabouts, that if pieces of amber were rubbed with cloth, the amber became capable of attracting small feathers, light bits of wool and so on. Amber is a glassy yellowish brown substance that is the fossilized resin of a long-extinct pine tree that once grew on the shores of the Baltic Sea. The Greeks (as well as other ancient peoples) used it for its ornamental value and the Greek name for amber was "elektron."

Other substances besides amber gained this attracting force when rubbed, but amber was the classic example. Therefore, when William Gilbert (the court physician of Queen Elizabeth I of England) studied the attracting force he suggested it be termed *electricity*. Eventually, people came to recognize the existence of an "electric fluid" which might stay put, as in amber, or might flow, as in a metal wire.

By the 1870's, scientists grew to think, more and more, that just as matter was composed of tiny particles (atoms), so the electric fluid must also be composed of tiny particles. The Irish physicist G. Johnstone Stoney suggested, in 1891, that the quantity of electricity present in each of these particles be called an *electron*. This was accepted and soon the name was applied to the particle itself.

In 1932, the American physicist C. D. Anderson discovered a particle just the size of the electron, but possessing an opposite kind of electricity. Whereas the electron carried negative electricity, the new particle carried positive electricity and was therefore named *positron*. (Actually, the *r* in positron is due to a false analogy with the *r* in electron. There is no *r* in positive so that a more logical name for the new particle would have been "positon.") For a while, there was a move afoot to change the name of the ordinary negative-electricity electron to *negatron*, but that never caught on.

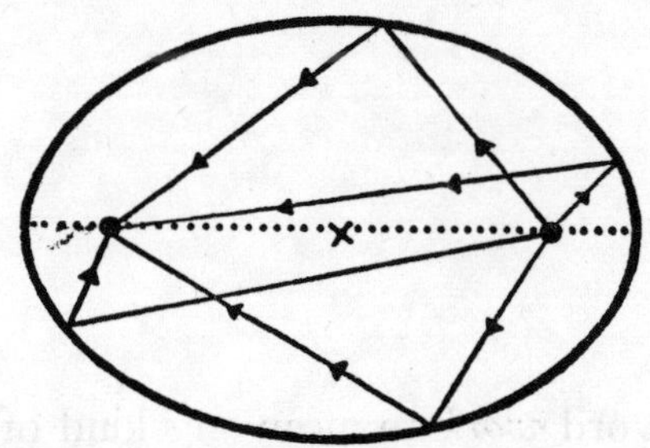

Ellipse

IF YOU were to hold up a *circle* (from the Latin "circulus" meaning "little ring") of cardboard to the light so that it cast a shadow on a smooth white surface, the shadow would be circular if the disk were held just between light and wall and just parallel to the line of the wall. If you were to tip the disk slightly, the shadow would no longer be exactly circular but would flatten somewhat into an oval. The more you tipped the disk, the flatter the oval.

This flattened circle is one of three related geometrical figures (see PARABOLA) studied by Apollonius about 250 B.C. He had a mathematical expression for each, and of the three the value of the expression was least for this flattened circle. Since the expression in this case was deficient compared to the others, and the Greek word for "deficient" is "elleipsis," the curve is called an *ellipse.*

There are two points within an ellipse, the *foci* (singular, *focus*). Imagine a large ellipse built out of mirrors facing inward. If a candle were placed at one focus of this ellipse, the rays of light traveling in all directions out from the candle would strike the mirrors at all points, and at every point the rays would be reflected in such a way that they would all converge upon the other focus.

Other curves also have foci, and lenses converge light to a focus. Different foci have different properties but the interest always lies in what happens to light rays emerging in all directions from a focus. Light rays also emerge in all directions from a fireplace and, as a matter of fact, "focus" in Latin, means "fireplace."

Halfway between the foci of an ellipse is its actual center. The flatter the ellipse, the farther apart are the foci and the farther each focus is from the center. Hence, the flatter the ellipse, the greater is its *eccentricity*, from the Greek "ek" (out of) and "kentron" (center). A circle, on the other hand, has its foci coinciding with its center so that it has an eccentricity of zero.

Energy

We all use the Anglo-Saxon word *work* to mean any kind of continued and purposeful activity, but to the physicist, the word has a more precise meaning. To him, work involves the movement of a body against a resisting force, and only that.

The idea of work involves two things: (1) the amount of push or pull required to force an object into motion against resistance (this push or pull is, appropriately, called *force*, from the Latin word "fortis," meaning "strong"); and (2) the distance through which the object is moved.

The Greek word for force is "dynamis," so physicists measure the amount of force by means of a unit called a *dyne*. For instance, two objects, each weighing 39 kilograms and separated by a distance of 10 centimeters attract one another with a gravitational force of one dyne.

If a force of one dyne moves a body through a distance of one centimeter, the amount of work done is one *dyne centimeter*. The dyne centimeter is also called the *erg* from the Greek word "ergon" meaning "work." For example, a man weighing 150 pounds (68,000 grams) lifting himself 8 feet (244 centimeters) against the resistance of gravity by climbing a flight of stairs does 68,000×244 or 16,600,000 ergs of work. (As you see, one erg is a very small amount of work.)

An object with the capacity to perform work (pent-up steam, a suspended rock, your own muscles, a taut bowstring, an atomic bomb) is said to contain *energy*. It has work ("ergon") in (Greek prefix, "en-") it. Energy can not only be turned into work, but work can be changed into energy.

The British physicist James P. Joule proved the latter by actual experiment in 1843. He showed that a fixed amount of work was always converted into a fixed amount of heat (a form of energy).

Since the erg, as I have said, is an inconveniently small unit, the amount of 10,000,000 ergs was set equal to one *joule* in Joule's honor. For everyday measurements, the joule is more convenient. Thus, the work involved in climbing a flight of stairs is 1.66 joules.

Engine

Originally, the word *engine* simply meant "an ingenious device." In fact the word *engine* is only a corruption of the word *ingenious*, which comes from the Latin "in-" (in) and "gignere" (to produce). Clever ideas, you see, are "produced in" the mind of an ingenious man. After James Watt invented a practical steam engine (an ingenious device run by steam power), the word was applied more and more to those devices particularly which took power from some nonliving source and turned it into work by means of the to-and-fro motion of a cylinder. These days, the gasoline engines that power our automobiles, buses, trucks, and aircraft are more important than the old steam engine.

The older, more general meaning of the word still persists in *cotton gin*, a machine used to strip the cotton fibers from the seeds. *Gin* is only a slangy contraction of *engine*.

Devices which turn power into work by turning rather than by piston motion are called *turbines*, from the Latin "turbo" (a top, or other spinning object). Water wheels, which are turned by running water and which produce power as a result, are, more properly, *water turbines*. There are also *steam turbines* where a jet of steam is the driving force, and *gas turbines* where burning gasoline or other fuel is the driving force.

A power-producing device which is turned by electricity is a *motor*, so called from the Latin "movere" (to move) because, after all, it is the part of the device that moves. My electric typewriter stays put while I work it but if I lift the top I can look in and see a furiously moving motor.

In popular speech, however, the word *motor* is also applied to any vehicle that moves a person from here to there by use of mechanical power, so that we speak of a motorcar or a motorboat, even though such objects are powered by engines rather than by motors.

Enzyme

In the early 1800's, the notion was beginning to arise that the body produced certain substances that brought about particular chemical changes useful to the body. For instance, stomach juices contained something that digested and liquefied meat. (The word *digest* comes from the Latin "digerere" meaning "to dissolve." The perfect passive participle is "digestum.")

At first, the digestive action was attributed to the hydrochloric acid contained in the juices, but, in 1835, the German physiologist Theodor Schwann reported that stomach juice contained something other than hydrochloric acid, and that this also had a digestive action. He called the new substance *pepsin*, from the Greek "pepsis" meaning both "cooking" and "digestion."

This news was greeted skeptically at first, but other such substances were found in saliva and in intestinal juices. This group of substances came to be known as *ferments* because their action seemed similar to that involved in the conversion of sugar and starch to alcohol (see FERMENT). In fact, by 1839, several people, including Schwann himself, proved yeast cells to be small living things, so that alcoholic fermentation was also associated with life.

For a while, some scientists thought there were two kinds of ferments. Those in digestive juices, found outside the living cell, were *unorganized ferments* and were chemicals, no more mysterious than hydrochloric acid, which also digested meat. The ferments in yeast which brought about alcoholic fermentation, however, were *organized ferments* and they involved a "life force" since those ferments were only found within cells. In 1878, the German physiologist Wilhelm Kühne applied the name *enzyme* to the unorganized ferments, from the Greek "en-" (in) and "zyme" (yeast), to show that they were similar in behavior to organized ferments in the yeast.

However, in 1897, the German chemist Eduard Buchner ground up yeast cells into a mash and filtered off the juice. The juice could still bring about fermentation, proving that neither intact cells nor "life force" were necessary. All ferments are basically alike whether inside or outside a cell, and all are now called enzymes.

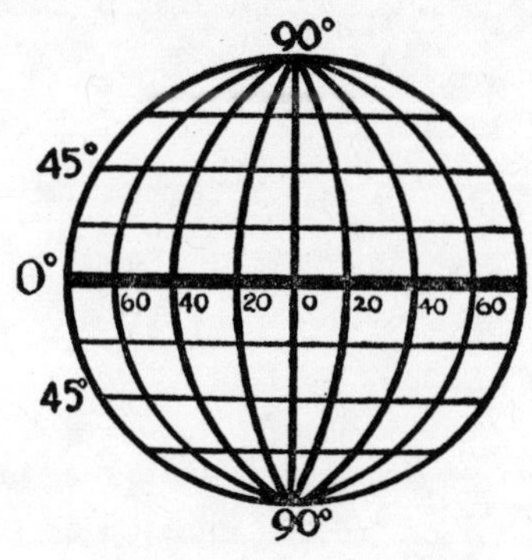

Equator

POSITIONS ON the surface of the earth can be described as being so many degrees east or west of the prime meridian (see MERIDIAN). This is the *longitude*, so called because the meridians by which it is measured are imaginary lines running north and south on the surface of the earth, or up and down on the conventional maps, so that they may be said to run "longways."

The prime meridian is 0 degrees, and measurements proceed both ways, so that there is a 10-degree *east longitude* and a 10-degree *west longitude*, for instance. The numbers increase until the series of meridians meet again directly opposite the prime meridian at 180 degrees.

In order to measure distance on the earth's surface north and south, another reference line, perpendicular to the prime meridian, must be used. The most natural such line is the *equator* which circles the earth in an east-west direction and is equidistant from the poles. A series of lines parallel to the equator can be run north and south all the way to the poles, and these are called, naturally, *parallels*. (The meridians are not parallel, but converge at the poles.)

The parallels define the *latitude* of a point, since they run east and west or "sideways" on the conventional maps, and the Latin word for "side" is "latus." Latitudes are counted both north and south so that the north pole is 90 degrees *north latitude* and the south pole 90 degrees *south latitude*.

Any line on the earth which (in imagination) cuts earth into two equal parts is a *great circle*, because it is the greatest circle that can be drawn on earth. All meridians are great circles, but the equator is the only parallel that is a great circle, so its name (from the Latin "aequator," meaning "one who equalizes") is appropriate. When the noonday sun is over the equator (see EQUINOX), day and night are equal in length and it is that equalization that probably gave the line its name originally.

Equinox

THE TILTING of the earth's axis causes the noonday sun to seem to climb higher and higher in the sky for six months, then sink lower and lower for the next six months (see SOLSTICE). At its most northerly high point, the sun is over the Tropic of Cancer at noon and the days are then longest (and the nights shortest) in the northern hemisphere and vice versa in the southern hemisphere. At its most southerly high point, the sun is over the Tropic of Capricorn at noon and the days are then shortest (and the nights longest) in the northern hemisphere and vice versa in the southern hemisphere.

As the noonday sun passes from its Tropic of Capricorn point, which it reaches on December 21, to its Tropic of Cancer point, which it reaches on June 21, it must pass, midway, the equator. It is over the equator on March 21, and at that time, day and night are equal in length everywhere on earth. This is the *vernal equinox*. *Equinox* comes from the Latin "aequus" (equal) and "nox" (night). It is the time of "equal nights." *Vernal* is from the Latin "vernalis," meaning "spring," since March 21 ushers in the spring of the year.

The noonday sun passes over the equator once again on its way south from the Tropic of Cancer back to the Tropic of Capricorn. September 22, which ushers in autumn, is the date of the *autumnal equinox*.

The earth bulges somewhat about the equator and the pull of sun and moon upon this bulge causes the earth's axis to twist about so that the North and South Poles trace a complete circle every 26,000 years. From earth's surface, it seems as though it is the vault of the skies itself that slowly twists. The result is that every year, the sun, when it crosses the equator at the time of the equinox, is seen in a slightly different spot in the sky, a little to the east of where it made the same crossing the year before. The point of crossing precedes (is earlier, is more eastward than) the last similar point of crossing, so that the 26,000-year circular movement of earth's axis is called the *precession of the equinoxes*.

Erosion

One of the ways in which the face of the earth is being continuously changed is through the action of water, falling from the skies as rain and rolling across the land to the sea, dragging soil and pieces of rock with it. The amount of water involved in this process amounts to about 8000 cubic miles each year so that a fair quantity of soil can be dragged along. In fact, enough soil is brought down to build flat areas miles out into the ocean (see DELTA). In the process, the river and its tributary streams cut out a bed for themselves in the rock and soil of the land. The process by which the energy of the flowing water cuts away a bed is called *erosion* from the Latin "e-" (away, off, out) and "rodere" (to gnaw). It is a "gnawing away" at the soil, in other words.

The amount of erosion depends upon the *gradient* of the river (from the Latin "gradus" meaning "step"); in other words, upon the steepness of the "steps" taken by the river from highland to lowland. The steeper the gradient, the faster the flow, the greater the erosion. The softness or hardness of the rock across which the river flows also counts. Sometimes a river will fall over a cliff of resistant rock that it has scarcely touched into an eroded section of soft rock largely eaten away. There results a *waterfall* (derivation obvious) or *cascade* (from the Italian "cascare," meaning "to fall," which in turn comes from the Latin "cadere").

Over relatively flat land, the river flows slowly, erodes little, and takes up an irregular winding course called *meandering* from the Meander River of Asia Minor, which does just that. On the other hand, under proper circumstances, a river like the Colorado can cut deeply into the land and form a *canyon.* This is a Spanish word (Spanish-speaking people settled the Colorado River territory before English-speaking people did) which comes originally from the Latin "canna" (a reed). We speak of the sugar *cane*, for instance, which is actually a reed. The gorge formed by a river is a long, hollow thing like a reed, and therefore *canyon.* The Colorado River forms the *Grand Canyon, grand* meaning "large" in most European languages.

Ether

BEFORE MODERN times, gases were not really understood. The very word *gas*, which is now used to represent any matter in an airlike state, was first invented about 1600 by J. B. van Helmont, a Flemish chemist, who got it from the word "chaos," the Greek term for the mysterious unformed material out of which the universe was made.

This almost superstitious awe of gases, which could be neither seen nor touched but which existed, was also shown in the way volatile liquids (those, that is, that turned easily into *vapor* — "vapor" being the Latin word for "steam") were named. They were referred to as *spirits*.

One such spirit, known since the 1200's, resulted from the action of sulfuric acid on alcohol (itself a spirit — and alcoholic beverages are still called *spirits* today). The new spirit was the most volatile liquid that had yet been discovered. If left standing, it vanished with extraordinary quickness. The chemist Frobenius named it *spiritus aethereus* in 1730.

The "aethereus" was a reference to the Greek "aither," their term for an imaginary and incorruptible substance supposed to fill the heavens. The volatile liquid Frobenius studied seemed so eager to leave our crude imperfect earth and depart for the heavenly realms above that it could be nothing but a "spirit of the aither" anxious to return home.

Eventually, however, the poetry of this thought was lost under the stress of practical usage (and better understanding of gases and vapors) and the name was shortened to a simple *ether*.

Nowadays, the term *ether* applies to a whole class of organic compounds with a structure similar to that of the original ether. The original ether was found to contain a pair of two-carbon groupings as part of its molecule. Such groupings were named *ethyl* groups, from "ether" and the Greek word "hyle," meaning "matter." The full name of the original ether is now *diethyl ether*, the prefix "di-" coming from the Greek word "dyo," meaning "two."

Evolution

THE FRENCH naturalist Georges Cuvier founded the science of comparative anatomy. Anatomy itself is the science that deals with the study of the physical structure of an organism. This structure can be properly studied only if the organism is carefully cut up so that its interior can be looked at. From the Greek "ana" (up) and "temnein" (to cut), the word *anatomy* ("to cut up") arises.

By *comparative anatomy* is meant the study that compares the anatomy of one creature with another to show relationships among them. Cuvier even compared the anatomy of existing creatures with those of extinct creatures as revealed by fossil remains (see FOSSIL). In this way, he showed that there were series of extinct creatures, existing through time, each a little different from the one before. He decided that there must be periodic catastrophes that wiped out life on earth and that after each, new and somewhat different life-forms were created.

Others, however, suggested that life was a continuous thing but that individual forms might, with time, slowly change into new and different species. Life would then be like a scroll, rolling out and revealing new and still newer life forms until out of very simple beginnings all the complex variety of modern life came into existence, while some forms of life, equally complex, might first have been formed and then have passed out of existence. This theory is termed *evolution*, from the Latin "e-" (out) and "volvere" (to roll); a "rolling out" (Latin, "evolutio"), in other words.

The English naturalist Charles Robert Darwin was not the first to think of evolution, by any means, but he collected so much evidence in its favor and, in 1859, published such an excellent book about it (*The Origin of Species*, which sold out its entire first printing on the first day of publication) that he might as well be considered the inventor of the theory.

It was only a generation later that biologists first learned of the method of change (see MUTATION) by which differences could be brought about between parent and offspring so that evolution might take place.

Ferment

One chemical discovery that was definitely prehistoric was the fact that fruit juice, if allowed to stand, changed in nature. The taste changed and the effects on the human being who drank it were odd and sometimes striking. (The fruit juice had become wine.) The change was accompanied by the formation of bubbles in the juice and this was the most noticeable outward sign of the change. (As we know today, the sugar in the juice was being converted to alcohol and to the gas, carbon dioxide.)

Similarly, a dough made of mashed grain moistened with water also begins to change if allowed to stand. (The starch is converted to alcohol and carbon dioxide.) Bubbles form and are trapped in the sticky dough, causing the whole mass to be raised up. If baked in the risen position, a light fluffy loaf of bread results (with the alcohol driven off by the heat), rather than a hard, compact pancake of baked flour. (The latter is just as nourishing but not as pleasant to eat.) A piece of the rising dough, if saved and added to a fresh batch, would hasten the change in the new batch.

The lump of rising dough is called *leaven* from the Latin "levare" (to raise). Bread made with it is *leavened bread*; without, *unleavened bread*.

This changing of fruit juice and moistened grain mixtures is called *fermentation* from the Latin "fermentare" (to cause to rise) which in turn comes from "fervere" (to boil) because bubbles appeared in the process as they did in ordinary boiling.

Exactly what it was that brought about the changes was not known until the invention of the microscope made it possible to see the tiny yeast cells that were responsible. The whatever-it-was, however, received names, all derived from the notion of boiling.

In Latin, it was called "fermentum," from which comes our word, *ferment*. In Sanskrit the word for "it boils" is "yasati" from which, perhaps, comes our word *yeast*.

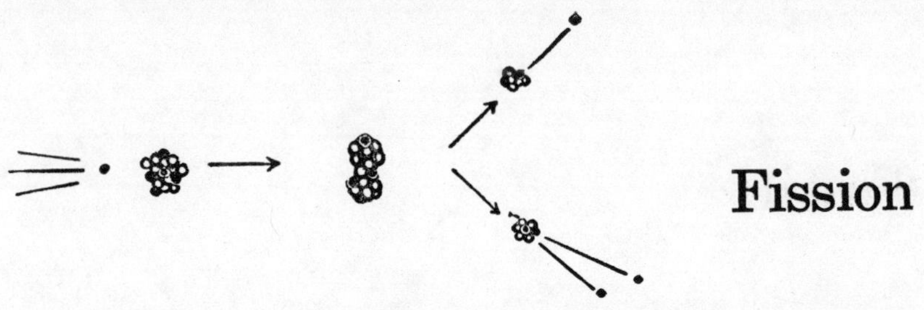

Fission

UNTIL 1939, the only nuclear reactions (see NUCLEUS) known were those involving rather minor changes in the nucleus — a rearrangement of particles or the loss of one to four of them. At most, a nucleus lost only about 1 to 1½ per cent of its mass in the form of small particles. It got so that physicists accepted this sort of thing and expected no more.

In 1934, the Italian physicist Enrico Fermi bombarded uranium with neutrons and found he got confusing results. Supposing that the uranium nucleus had been changed only slightly, he tried to explain his results on that basis and just got entangled. Others who repeated the experiments had no better luck.

In 1938, the German physicists Otto Hahn and Fritz Strassman finally decided that they just had to believe their own chemical testing. They had barium mixed in with the bombarded uranium, although the barium atom was so much smaller than the uranium atom, they could not see how it could have come there. Later that year, a German physicist (in exile), Lise Meitner, suggested that when the neutron hit the uranium nucleus, it split that nucleus into two nearly equal parts, one of the parts being barium.

It was unheard of; her theory created a sensation. Physicists all over the world (and in America especially) began checking and she was right! The uranium nucleus did break in two! This new nuclear reaction was called *fission* from the Latin "fissio" (infinitive form, "findere"), meaning "split." The nucleus, after all, was not merely chipped; it was split in two.

Fission gives off several times as much energy as ordinary nuclear reactions and, besides, liberates neutrons that can cause neighboring atoms to undergo fission. Each atom-split causes more atom-splits which proceed like links in a chain. This is called a *chain reaction.*

This chain reaction made it possible to develop a nuclear reaction that would be self-sustaining (i.e., would continue of its own accord, so to speak, once started; just as wood burns of its own accord once started). As it happened, the first such self-sustaining fission was achieved under the leadership of Fermi, who started it all.

Fossil

THE HISTORY of the earth during the billions of years of its existence has been deduced from a study of the rock formations of its crust and, in particular, from the remains of once-living organisms which have been dug out of the ground. These remains (called *fossils*, from the Latin "fossilis," from "fodere," meaning "to dig") were observed long before modern times but for many centuries were considered to be ordinary rocks that accidentally resembled living things or, perhaps, the remnants of animals drowned in Noah's flood.

In 1791, however, an English land-surveyor, William Smith, showed that different rock layers contained different types of fossil, and that a given layer could be traced across broken ground by following the location of its characteristic fossils.

The French anatomist Georges Cuvier studied the anatomies of these fossil remains in about 1796 and began to show that as much sense could be made out of them as if they represented living animals. Furthermore, he showed that some fossils represent animals completely different from any existing today.

Because of the importance of fossils, earth's history was divided into broad stretches of time according to the nature of the life then existing. At first, the three major divisions, counting backward, were the *Cenozoic*, the *Mesozoic*, and the *Paleozoic* eras. The suffix "-zoic" comes from the Greek "zoon" (animal). The various prefixes are from the Greek "kainos" (new), "mesos" (middle), and "palaios" (old). So there was the era of "new animals," with mammals dominant, in the last 60 million years; of "middle animals," with reptiles dominant, in the 100 million years before that; and of "old animals," with fish and land invertebrates dominant, in the 300 million years before that.

Later, earlier eras were established: the *Proterozoic*, with sea invertebrates dominant (from the Greek "proteros," meaning "earlier"), and the *Archeozoic*, with only one-celled animals existing (from the Greek "archaios," meaning "ancient").

Fraction

EVERY CHILD begins his study of mathematics with a consideration of the whole numbers: 1, 2, 3, and so on. So did mankind as a group. Slowly, and in steps, men manipulated digits and came up against unexpected problems that taught them about numbers other than digits. For instance, there are two two's in four, three two's in six and so on. In other words, $4/2$ (four divided by two) is 2 and $6/2$ (six divided by two) is 3.

But how many two's are there in five? More than two, certainly, but less than three. In this way, men were forced to think of numbers lying between digits. The quantity $5/2$ (five divided by two) can only be two and a half; that is, one unit plus one unit plus half a unit. Half a unit can be expressed logically as $1/2$ (one divided by two).

Because a number like $1/2$ represents a unit broken into two equal parts, it, and numbers like it, are termed *fractions* from the Latin "fractus" (broken).

By using fractions, one can locate numbers at various points between digits. The midpoint between 1 and 0 is $1/2$, of course. The midpoint between $1/2$ and 0 is $1/4$ while the midpoint between $1/2$ and 1 is $3/4$. You can also locate $1/3$, $2/3$, and so on. Between 5 and 6 you can do the same by having $5\,1/2$, $5\,1/4$, $5\,3/4$, $5\,1/3$, $5\,2/3$, and so on.

Such numbers, which seem to fill up all the space between the digits, are obtained, you see, by comparing two digits. The figure "two-thirds" is obtained by comparing 2 and 3. The digit 2 is two-thirds as large as 3, hence $2/3$ means "two-thirds." Such numbers seem to involve a mental comparison of digits, thought, and judgment. The Latin word for "accounting" or "reckoning" is "ratio." Two numbers put side by side and compared form a *ratio*.

Any number which can be expressed as the ratio of two digits is a *rational number*. Such numbers include digits as well as ordinary fractions since 5, for instance, can be written as the ratio $5/1$ or $10/2$.

Friction

Isaac Newton's First Law of Motion states that a moving object will continue moving in a straight line forever unless acted on by some external force that will serve to slow it, speed it, or change its direction. It took a long time to discover this truth since in actual fact there always exist external forces which are easy to overlook.

For instance, slide a hockey puck along the ice. It will move a long way in a straight line, but it will slow constantly and finally come to a halt. If Newton's Law is correct, why should that happen? Nothing seemed to touch it or affect it in any way.

Actually, all the time the puck is moving, it is rubbing against the ice. There are tiny irregularities in the puck's lower surface and tiny irregularities in the ice. These catch at one another as the two surfaces rub together and this absorbs some of the puck's energy of motion, so that the puck slows. If the puck were sliding along wood, which is less smooth than ice, the puck would stop sooner. If it were sliding along brick, it would stop still sooner.

This slowing force resulting from the rubbing of uneven surfaces is called *friction*, from the Latin "fricare" (to rub).

Even if motion does not involve one solid rubbing against another, there are forces slowing motion. A ship moving through water has to force molecules of water apart and must overcome the attraction between the water molecules. This has the same effect as friction. Just as one solid may be rougher, hence display more friction than another (as brick is rougher than ice), so one liquid may be "thicker" and harder to move through than another (as molasses is "thicker" than water). This "thickness" is referred to as *viscosity*, from the Latin "viscum," a word for a kind of thick and sticky birdlime.

Even gases have viscosity, and a great deal of airplane design involves itself with methods for reducing air resistance (called "drag" for obvious reasons) so that as little power as possible is wasted.

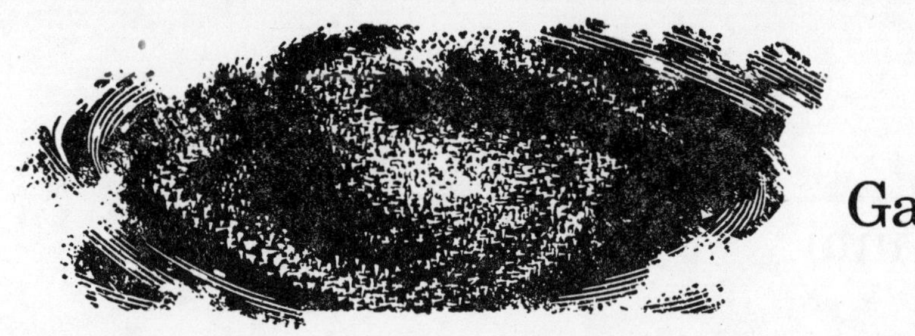

Galaxy

THE SUN IS one of a gigantic cluster of perhaps a hundred billion stars, arranged in lens form. The diameter of this lens is about three times its thickness. We are near the outer, thin end of the lens.

The stars near our sun are seen as individual points of light and, looking through the thickness of the lens, we see to the emptiness beyond. However, if we look through the long diameter of the lens, the stars dim with distance and are too numerous to be seen through. They form a soft luminous band that encircles the sky.

The Greeks thought of it as a belt of milk in the sky and called it "galaxias" from their word "gala," meaning "milk," and today we call the entire cluster of stars of which we are part the *Galaxy*.

The Latins called it "via lactea," meaning "road of milk," and we call it just that ourselves — the *Milky Way*.

In the early days of telescopic observations, a number of cloudy objects were detected among the stars, particularly by the French astronomer Charles Messier, by Sir William Herschel (see URANIUM) and his son, John Herschel. These objects were called *nebulae* (singular, *nebula*), which meant "clouds" in Latin and came from the Greek word "nephele," meaning "cloud." Many of the nebulae turned out to be patches of dust within the Galaxy, seen as black clouds because they blotted out sections of the Milky Way behind them, or as illuminated clouds, lit by stars within them. A number, however, like the Andromeda Nebula, turned out to lie far outside the Galaxy and, indeed, to be collections of stars as large as the Galaxy, appearing small only because of their tremendous distances.

These outside nebulae are called *extragalactic nebulae* ("extra" being a Latin prefix meaning "outside") to distinguish them from the relatively minor dust-patches inside the Galaxy. There are hundreds of billions of such extragalactic nebulae in existence. Sometimes they are referred to, poetically, as *island universes*, but more and more the term *galaxy* is being applied to all of them so that we can speak of galaxies in the plural. In fact, because the galaxies themselves exist in clusters, the phrase "galaxy of galaxies" is now being used.

Gallium

THE DISCOVERY of an element affords an excellent chance for its discoverer to combine science and patriotism by naming the element after his native country. The most recent case was that of element 95 which was first prepared in 1944 by a group of American chemists under the direction of Glenn T. Seaborg. The new element was named *americium.*

Earlier, in 1939, the French chemist Marguerite Perey detected element 87, and eventually (in 1946) named it *francium* after her native country.

Still earlier, in 1898, Pierre and Marie Curie, in their researches on a uranium ore, had located small amounts of element 84. (Later that same year they were to make their famous discovery of radium, — see RADIOACTIVITY — but that was the second element they discovered, element 84 being the first.) Madame Curie was a Pole by birth, her maiden name being Marie Sklodowska, so the element was named *polonium* after her native country.

And still earlier, in 1886, the German chemist Clemens Alexander Winkler discovered element 32, and named it *germanium* after his native country. (Of course, to a German, the name of his country is Deutschland, but Winkler used the Latin name of "Germania.")

The most interesting case is the earliest and it involves a certain point of ethics. Elements have been named after individuals, but it has been customary to name them only after dead ones. For instance, elements 99 and 100, first discovered in 1955, were named *einsteinium* and *fermium* after Albert Einstein and Enrico Fermi, who both richly deserved the honor and who had both recently died.

In 1875, however, the French chemist Lecoq de Boisbaudran discovered element 31 and named it *gallium.* The usual explanation of the name is that it is derived from Gallia, the Latin name of his native country. However, it has been pointed out that Lecoq (meaning "the rooster" in French)is, in Latin, "gallus." There is therefore the strong suspicion that de Boisbaudran successfully defied ethics by naming an element after not merely a living person, but after himself.

Gamma Globulin

Scientists often have to deal with substances that belong to the same family or group but are different in little ways. In naming them, it is often convenient, at first, to give them numbers or letters. To vary the monotony, Greek letters are sometimes used. The first three letters of the Greek alphabet are α, β, and γ. The names of these, spelled out in ordinary Roman letters, are alpha, beta and gamma. (From the first two, we get the word *alphabet.*)

An example of the use of these Greek letters involves the proteins of blood. The first protein to be obtained from blood came from the red corpuscles and was called globulin (see HEMOGLOBIN). This protein differed from egg albumin, the protein of egg white (see PROTEIN) and the one best known, by having a larger molecule and being less soluble in water.

Following the lead of Felix Hoppe-Seyler, a German physiologist, it became customary to divide the simpler proteins into two groups: the smaller, more soluble albumins; and the larger, less soluble globulins. (Oddly enough, the protein of the red corpuscles, the original *globulin*, turned out to be a poor representative of the group, is now called *globin*, and is not considered to be a true globulin.)

The liquid part of blood (the plasma) contains proteins of both the albumin and globulin variety dissolved in it. These proteins can be separated by subjecting the plasma to the influence of an electric field. The various proteins in it move under the influence but at different rates, so that they gradually separate. The albumins move more quickly than the globulins and stick together. The globulins break up into three groups. And now we have the example of Greek-letter-naming I was going to give you. The fastest-moving globulins were named *alpha globulins*, the next *beta globulins*, and the slowest *gamma globulins*.

The gamma globulins turned out to be the headline-makers, since they include chemicals that govern the body's resistance to many diseases.

Gene

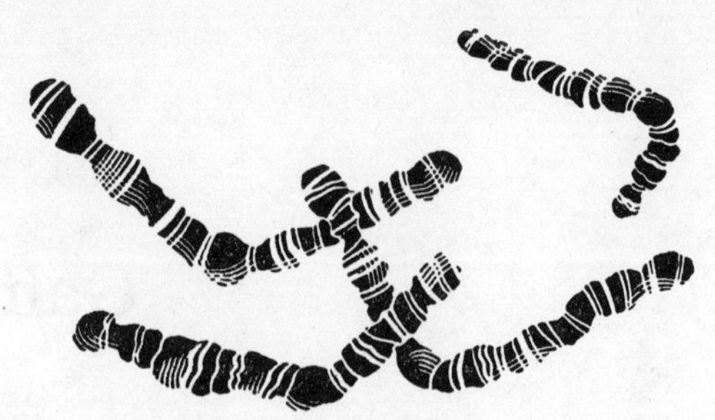

One of the first ways in which biologists tried to find out something about the inside of the cell was to subject it to the action of various dyes. The different substances within the cell reacted differently, as was to be expected, so that some objects would show up colored against a colorless background (or vice versa). Within the nucleus of the cell, for instance, there were certain small areas that would bind dye tightly and show up colored. In 1879, the German anatomist Walter Fleming, therefore, called these bits *chromatin,* from the Greek "chroma" (color).

If this method of staining is used on cells in all stages of cell division, it turns out that at one stage during the process, the chromatin collects itself into small threadlike bodies. These threads of chromatin were called *chromosomes,* or "colored bodies," since the Greek "soma" means "body." Furthermore, since these threads of chromatin play such a noticeable part in cell division, that process was given the name *mitosis,* from the Greek "mitos" (thread).

It became quite apparent that the chromosomes somehow directed the chemistry of the body and that children, because they inherited half their chromosomes from their mother and half from their father, had some characteristics resembling those of the family of each parent. Since human cells contain 46 chromosomes while each individual is made up of thousands of inherited traits, biologists have assumed that each chromosome is made up of hundreds of smaller units, each of which controls one characteristic. These smaller units they call *genes* from the Greek suffix "-genes" (giving birth to) — see OXYGEN, HYDROGEN, NITROGEN — which in turn arises from the Greek "gignesthai" ("to be born" or "to be produced").

From the word *gene* comes the name for the science that makes a study of how traits are inherited — *genetics.* The study of how to encourage the inheritance of valuable traits (a study that is still in its infancy, but which many pseudo-scientists and semieducated people love to talk about) is *eugenics,* the Greek prefix "eu-" meaning "well."

Glucose

In the early days of chemistry, when modern instrumentation was absent, the tongue had to substitute as one means of distinguishing chemicals. In this way, the class of sour substances (see ACID) was identified.

In the same way, sweet substances were noted. The Greek word for "sweet" is "glykys" and so a particular variety of sugar (there are a number of chemical varieties) is now named *glucose.* (The "-ose" suffix is now commonly used by chemists for sugars and related compounds.) As it happens, glucose is not as sweet as ordinary table sugar.

Glucose is found in small concentration in blood and is the immediate source of the body's energy so that it is sometimes called *blood sugar.* In 1857, the French physiologist Claude Bernard discovered in liver a starchy substance which the body could convert to glucose at need. It was called *glycogen,* the Greek suffix "-genes" meaning "produce." Glycogen was the "producer of sweetness."

Other chemical compounds which tasted sweet received similar names. An organic liquid which is actually sweeter than table sugar to the taste, but fairly poisonous and used mainly as an antifreeze, is called *glycol.* (The "-ol" ending is used by chemists for alcohol and related compounds.) A similar compound, slightly more complicated, which makes up part of the molecules of oil and fat and is quite harmless (in fact, it is used in candies), is as sweet as table sugar and is called *glycerol.*

Glucose, glycol, and glycerol are all chemically similar in that they contain within their molecule atom-groups made up of an oxygen atom and a hydrogen atom, such a combination being called a *hydroxyl group.* Nevertheless, molecules without the hydroxyl group, such as the amino acid, glycine, can be sweet and glycine is named for that fact (see GLYCINE). (The "-ine" ending is mostly reserved by chemists for nitrogen-containing organic compounds.)

Even the element beryllium has the alternate name of *glucinum* because some of its compounds are supposed to taste sweet.

Glycine

The thickening, sticky material in dough is *gluten* and from it comes our word *glue* for anything thick and sticky. Important sources of glue are the hides, hoofs, and bones of animals. The substance in these that is responsible for the gluiness of the final product is a fibrous protein named *collagen*, from the Greek "kolla" (glue) and "-gen" (the suffix meaning "produces"). Collagen, in other words, "produces glue."

If pure collagen is made to undergo prolonged heating, the protein molecule breaks up somewhat to form *gelatin*. The word comes from the Latin "gelare" (to freeze), since a warm liquid solution of gelatin, if allowed to cool, "freezes" into a jelly. (The word *jelly* also comes from "gelare," by the way.)

In 1820, the French chemist H. Braconnet investigated gelatin chemically. Earlier it had been shown that cellulose (see PROTOPLASM), on treatment with acid, decomposed into simpler molecules that were a kind of sugar. Would gelatin, which came from collagen, a fibrous constituent of animals, do the same?

Gelatin did indeed break down into smaller fragments on treatment with acid, and at least one of these fragments, when purified, was sweet to the taste. Braconnet was sure it was a sugar and called it, forthrightly, "sugar of gelatin." It wasn't till 1838 that "sugar of gelatin" was found to contain nitrogen, which ordinary sugars do not. It was renamed *glycine* from the Greek "glykys" (sweet). An alternate name, not much used, is *glycocoll*, meaning "sweet glue."

Braconnet had done more than he knew. Glycine was the simplest amino acid (see AMMONIA); not the first discovered, but the first to be shown to form part of a protein molecule, a milestone in biochemistry.

And it also turns out, derivationally speaking, that though the story of glycine starts out with glue, the coincidence of the first two letters has no significance.

Grammar

GRAMMAR IS the science that deals with the correct (or, at least, accepted) use of language. It is derived from the Greek "gramma" (letter), the verbal form of which is "graphein" (to write). In ancient and medieval times, grammar was one of the most important branches of learning (see LIBERAL ARTS) but in modern times it has been crowded by the advance of the physical sciences.

Yet in one respect, it resembles some of these physical sciences. The words with which the science of grammar deals are as carefully classified as are the animals with which zoology deals or the elements with which chemistry deals.

For instance, there are *nouns* — words that name persons or things — "John," "rat," "table." The word *noun* comes from the Latin "nomen" meaning "name."

A *pronoun* is also a name, but one for an object already mentioned by a specific name. The pronoun is not specific: "I," "which," "who." One of the meanings of the Latin "pro-" is "substituted for," so a pronoun is a word "substituted for a noun."

A *verb* denotes an action ("to smash"), a state ("to be") or a happening ("to age"). These are the words which can least be spared, perhaps. Certainly, the derivation would seem to make them most important since the Latin "verbum" means simply "word."

An *adverb* explains the nature of a verb (walk "slowly"; insist "stubbornly"). The Latin prefix "ad-" means "to"; so an adverb is added "to a verb" to explain it.

Similarly, an *adjective* describes or defines a noun ("white" hat; "fast" train). It is derived from the Latin "ad-" (to) and "jacere" (to throw). The noun and adjective are "thrown together," so to speak.

One more example is the *interjection*, an exclamation that intrudes on ordinary speech ("Ouch," "Oh"), from the Latin "inter" (between) and "jacere" (to throw). It is a word "thrown between" ordinary words or sentences.

Granite

THE ROCK which makes up most of the dry land on earth is granite. This is, in turn, made up of a mixture of three different kinds of rock: mica, feldspar, and quartz.

Mica can be easily split into thin transparent sheets and has therefore been used to form windows on stoves where it is necessary to have something transparent that won't burn or melt. The thin sheets are glossy and shiny so that the name may come from the Latin "micare" (to shine).

Feldspar is the most common rock about us. It makes up perhaps 60 per cent of the dry land, and the name shows that. *Feld* is the German word for "field" while *spar* is from an Anglo-Saxon word for any rock that is not metal-containing. (A rock that is metal-containing is an *ore*. This word is much more familiar to us because ores are rarer, more useful, and more valuable than spars, so that the ores are more spoken of.) Feldspar is thus "field rock," the common rock of the fields.

Quartz is from a German word, *Quarz*, which is of uncertain origin. It is also very common and, when broken into small fragments by the action of wind and water, it is *sand* (a word of Anglo-Saxon origin).

Granite is an igneous rock (see IGNEOUS) and when it forms by cooling, the three components, mica, feldspar, and quartz, form crystals large enough to be seen separately. Granite, therefore, does not present a smooth and homogeneous appearance but is obviously made up of mixed grains of different substances. The Latin word for "grain" is "granum" and so *granite* is the "grained rock," so to speak.

An even more common igneous rock is *basalt*, which underlies the granite of the continents and makes up most of the earth's crust under the oceans. It is darker and heavier than granite. According to Pliny (Gaius Plinius Secundus) the Roman naturalist, the word originated in Ethiopia, where it was applied to a dark variety of marble. The meaning later spread to include any dark or blackish rock, particularly the one we call basalt today.

Gravity

Isaac Newton's First Law of Motion states that every object in motion will move in a straight line unless forced to change direction by some external push or pull. If you were whirling in a circle at a great speed, held only by the handle of an elastic rope in your clenched fist, and were to relax your hold at any time, you would fly off instantly out of the circle in which you had previously been moving. It is the constant pull of the rope that forces you to change the direction of your motion constantly.

The force tending to move you out of the circle is *centrifugal* force, from the Latin "centrum" (center) and "fugere" (to flee). It is the force that causes you to flee the center. The force pulling you toward the center (the pull of the elastic rope) is a *centripetal* force, from the Latin "petere" (to move toward). It is the force that causes you to move toward the center. The balance of the two keeps you in the circle.

The rope keeps you from flying away because it is held strongly together by molecular cohesion (see CAPILLARITY). When centripetal force is the result of such cohesion, it works only if there is a continuous material connection between yourself and the center.

All matter, however, attracts all other matter, even without physical contact. If the lump of matter is large enough, the attraction is considerable. For instance, an object released in space a thousand miles from earth is attracted by the earth and moves toward it, although there is a vacuum between. The moon, 237,000 miles away, is also falling, but it has a motion of its own and centrifugal force balances its falling motion, so that it stays in a closed orbit about the earth, never falling altogether and never escaping altogether.

This pull without physical contact is what gives us the sensation of weight or heaviness. The Latin word for "heavy" is "gravis" and so the attraction of one body for another is *gravity* and that is the centripetal force that holds the solar system together and everything on earth in place.

Helium

In 1868, a total eclipse was visible in India and, for the first time, the sun's atmosphere (best observed during eclipses) could be studied by the new technique of spectroscopic analysis (see SPECTRUM). This had been developed only nine years earlier and consisted of passing the light radiated from a white-hot substance through a glass prism. The light is split up into lines of different colors, and each element forms its own characteristic pattern of colored lines in fixed positions.

The French astronomer Pierre J. C. Janssen allowed the light of the solar atmosphere to pass through the prism during the Indian eclipse and noticed that among the familiar lines of earthly substances, a yellow line was produced which he could not identify. The British astronomer Sir Norman Lockyer compared the position of this line with those of similar lines produced by various elements, and decided that this new line was produced by an element in the sun that was not present, or had not yet been discovered, on earth. He called it *helium*, from the Greek word for the sun, "helios."

For decades that was how matters stood. Helium remained an oddly colored line in sunlight and nothing more. Few chemists took it seriously.

In 1888, the American chemist William F. Hillebrand, found that a uranium mineral named uraninite, when treated with strong acid, gave off bubbles of gas. He studied this and decided it was nitrogen. To be sure, some of the gas was nitrogen, but Hillebrand unfortunately ignored the fact that, when heated, some of its spectrum lines were not those of nitrogen.

The Scottish chemist Sir William Ramsay read of this experiment and was dissatisfied. He used another uranium mineral, cleveite, and, in 1895, repeated the experiment. He and Lockyer studied the spectral lines of the gas, and almost at once they realized what they had. Fully 27 years after helium had been discovered in the sun, it was finally located on earth.

Hemoglobin

ANTON VAN LEEUWENHOEK, the Dutchman who first used a microscope systematically (see MICROBE), was the first to see the small objects in blood which contain the red coloring matter. He called them *corpuscles*, from the Latin "corpusculum" (little body). Today, they are generally called *red corpuscles* to distinguish them from other little bodies in blood which do not contain coloring matter and are the *white corpuscles*.

Sometimes these are referred to as *red cells* and *white cells*, but in the case of the former, this is a misnomer. The white corpuscles are actually cells, but the red corpuscles are not. The red corpuscles lack a nucleus (see PROTOPLASM), which all true cells must have. And yet the name most used by scientists keeps this misnomer, since a red cell is most often called an *erythrocyte* from the Greek "erythros" (red) and "kytos" (a hollow space or cell). Similarly, the white corpuscle, with greater justification, is a *leukocyte*, from the Greek "leukos" (white).

The red corpuscles were also called *globules* by van Leeuwenhoek from the Latin "globulus" (little ball). This, too, is a misnomer. Van Leeuwenhoek's early microscopes could not show him the exact shape, but we know now that the red corpuscles are not globes or spheres, but are disk-shaped, with a depression on each flat side (like a candy Life Saver with the hole not quite punched through). In fact, the red corpuscles are sometimes called *red disks* for this reason.

Nevertheless, in 1805 or thereabouts, when the Swedish chemist Jöns J. Berzelius obtained a colorless protein from the red corpuscles, he named it *globulin* because he got it from the "globules." Inside the corpuscles, the globulin was associated with the coloring matter, then called "hematin" (see PORPHYRIN). The combined molecule was named *hematoglobulin.* This, in time (even scientists are lazy), got shortened to *hemoglobin* by omitting four letters and that, today, is the name for the red protein of blood.

Hemophilia

In the last half-century, two royal families were afflicted with a disease marked by excessive bleeding. The son of Czar Nicholas II of Russia was a "bleeder." A small scratch was enough to put him in danger of bleeding to death. This was also true of certain sons of the Spanish king, Alfonso XIII. This received so much attention in the newspapers that the condition became known as the "royal disease." Its proper name is *hemophilia*, from the Greek words "haima" (blood) and "philia" (love), since the patient apparently "loved to bleed."

But hemophilia hit royalty only in modern times. The Spanish bleeders could be traced back to Queen Victoria of England through her youngest daughter, Beatrice, while the Russian bleeders could be traced back to the same Queen through her second daughter, Alice. It seems fairly certain that the "royal disease" originated with her, at least as far as royal families are concerned.

People suffer from hemophilia when they are born without a certain one of the numerous substances in blood that contribute to clotting. The disease is tied up with the sex of a person in such a way that while men can have the disease, they can't pass it on to their children; whereas women may not show it and yet be able to pass it on. Queen Victoria gave birth to at least two daughters who could transmit this failure of blood-clotting mechanism and who did, without themselves showing it, of course.

Actually, many perfectly nonroyal people have the disease, or other similar bleeding diseases brought on by lack of clotting factors in the blood. One odd variety was first described in 1952. It occurred in a five-year-old boy whose last name was Christmas. The missing factor was therefore named *Christmas factor*, and this particular variety of bleeding condition was, with peculiar insensitivity, named *Christmas disease*.

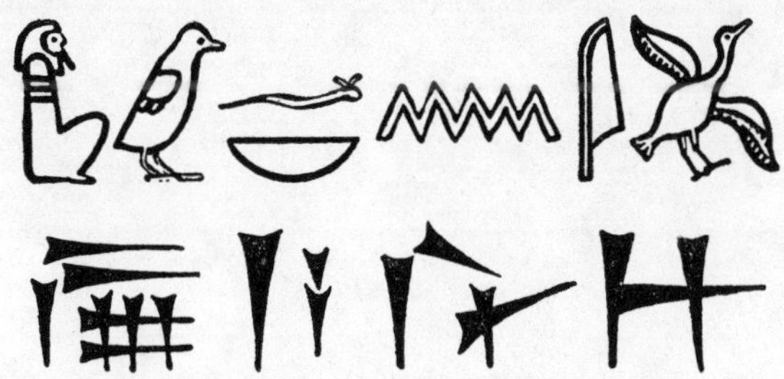

Hieroglyphic

LANGUAGES associated with a religion persist there even after they go out of ordinary use. Thus, Latin has always been the language of the Roman Catholic Church and Hebrew of the Judaic ritual, although both are usually dismissed as "dead languages."

This sort of thing was observable in ancient times, too. Greek tourists of the centuries before Christ, in an Egypt that was already ancient, found monuments with carved figures on them that ordinary Egyptians could no longer read. Only the priests preserved the language. The Greeks therefore called these figures "hieroglyphika" (English, *hieroglyphics*), from "hieros" (sacred) and "glyphein" (to carve); they were "sacred carvings."

After about 600 B.C., the Egyptians as a whole used a much less elaborate writing for ordinary work and this was called by the Greeks "demotika" (English, *demotic*) from "demos" (people). It was the writing of the common people.

The only other written language as early as (or possibly earlier than) the Egyptian, was the Sumerian. The Sumerians lived in the Tigris-Euphrates area (modern Iraq) and in the absence of stone inscribed their writings on clay with slanted jabs of a stylus. The Babylonians, Assyrians, and Persians all adopted this style of writing and it did not die out until 300 B.C. Because the marks on clay were wedge-shaped (▼), the writing was called *cuneiform* from the Latin "cuneus" (wedge) and "forma" (form). It was writing in the "form of a wedge."

These early writings were not alphabetical. Each different sign represented a separate word or idea. (Eventually, they reached the point where they represented only syllables and sometimes even letters, but this was a late development after the Phoenicians had pointed the way with a true alphabet.) Such "idea" signs are *ideographs. Idea* itself is a Greek word meaning "form," "shape" or "resemblance to reality," and "graphein" means "to write." An *ideograph* is an "idea writing." The best known example of a modern ideographic language is, of course, Chinese.

Hormone

THE LATIN WORD "glans" means "acorn." People have always been impressed by the smallness of acorns (probably because of the contrast of the giant oak which is the final result) and small lumps of tissue in the body were, therefore, called *glands* by the early anatomists (see LYMPH).

Such lumps are to be found over the kidney. The Latin word for "kidneys" is "renes" and the glands are the *suprarenals* or the *adrenals*. (The Latin "supra" means "above" and "ad" means "on" so the two names are pretty much the same.)

In 1895, the biochemists George Oliver and Edward Albert Sharpey-Schafer found that there was something in the adrenal glands which caused a contraction of the arteries and a rise in blood pressure. The actual substance that did this was isolated in 1901 by the Japanese biochemist Jokiche Takamine.

This was the first example of a chemical, formed in a gland and liberated in small quantities into the blood stream, that could excite specific organs to specific activities. In 1902, the English physiologists William M. Bayliss and E. H. Starling suggested that such chemicals be known as *hormones* from the Greek "horman," meaning "to set in motion."

Chemicals, and fluids generally, formed by small organs have become so important to physiologists that the word *gland* has come to mean any bodily organ that secretes a fluid, regardless of size. The liver is a gland though it weighs several pounds. On the other hand, the lymph glands, perhaps the first to be so named, should no longer really be called glands since they secrete no fluid. They are better termed lymph *nodes*, from the Latin "nodus" (knot), since they resemble knotlike swellings in the stringlike lymph vessels (see LYMPH).

As for the first hormone, the one from the suprarenals, it is called *adrenalin* from, as stated above, the Latin "ad" (on) and "renes" kidneys; or the equivalent name *epinephrine*, from the Greek "epi" (on) and "nephros" (kidney).

Humor

THE LIVER JUICE, which may be called either bile or gall (see BILE), has also a Greek name, "chole," which comes from their word "cholos" (bitter), and bile is bitter, indeed. (We have a common simile, "bitter as gall.")

The Greek word is used in English only in combination. For instance, the ancient Greeks thought there were four chief fluids in the body: blood, phlegm, yellow bile, and black bile. (Actually, there is only one bile, but when freshly formed it is golden yellow, whereas after standing awhile chemical changes take place that turn it a greenish black, and this is what may have given the Greeks their notion of two biles.)

If any one of these fluids was present in too great a quantity, thought the Greeks, the person suffered from immoderation in one way or another. If there was too much yellow bile, for instance, he had too great a tendency to anger and was *choleric* (the Greek "chole" showing up, you see). If he had an overbalance of black bile, he was too given to sadness and was *melancholic* (the Greek "melas" means "black," hence *melancholy* is "black bile").

In later times, these fluids were referred to as *humors* from the Latin word "humere" (to be moist). The theory of humors is gone now, but the word still lingers for any pronounced type of temperament that is changed from the normal. A person may be in a "bad humor" or a "good humor" or a "contrary humor," and so on.

In Elizabethan times, there was quite a fad for the writing of plays about people who had some pronounced one-track personality. One might be of a boastful "humor," one cowardly, one avaricious. Ben Jonson went to the deliberate extreme of a play that included only such characters, called *Every Man in his Humor*. Such plays were generally comedies and the characters, each harping away at his particular personality trait, were laugh-provoking. For that reason, the word *humorous* has come to mean "funny."

Hurricane

MOST STORMS are cyclonic in character (see CYCLONE) and in general are mild enough. However, every once in a while, conditions are such as to make the whirling motion of a cyclone too rapid for comfort.

An all-too-familiar condition to inhabitants of the eastern and Gulf seaboards of the United States is a cyclone that begins over the Caribbean in late summer or early fall, forms a gigantic whirlwind with winds of over a hundred miles an hour, and starts moving northwestward. This is called a *hurricane* from a Caribbean Indian word, "Hurakan," which was the name of one of their evil spirits. Anyone who has lived through a hurricane (as I have) will testify that the evil-spirit theory has its points.

Hurricanes (at least by name) are confined to the Atlantic. There are, however, similar severe cyclonic storms in the west Pacific. These are *typhoons*. This comes from the Arabic "tufan." Why Arabic? Well, the Arabs explored the southeastern reaches of Asia before Europeans did, so that there are strong Moslem groups in Indonesia and even in the Philippines to this day. (Some tribes in the southern Philippines are called Moros from the Spanish word meaning "Moors" because of their religion.) The Arabic word probably comes, in turn, from the Greek "Typhon" (or from some common ancestor), the name of an evil giant who fought with Zeus in a battle of storm and lightning. (Again the evil-spirit theory.)

Sometimes, over land surfaces, much smaller but more intense cyclones are set up, with winds so fierce (their speed has never been measured since they destroy all measuring devices) that they devastate whatever they pass over. Fortunately, their path is short and narrow. These are *tornadoes*. The Spanish word "tornado" means "return" (the wind moves and returns in a circular path over and over), but it has also been derived from the Spanish "tronada" (thunderstorm). Why Spanish? Well, the Spaniards colonized the southwestern and middle areas of the United States before the English-speaking Americans arrived and they experienced these storms (which breed best in the south central U.S.) before we did.

Hydrogen

The British chemist Henry Cavendish was the first (in 1766) to study systematically a gas he obtained by treating iron filings with acid. Because the gas burned when heated he called it "inflammable air from the metals."

To the early chemists, still more remarkable than the mere fact that the gas burned was that, after burning, it left behind a liquid substance that proved to be pure water. Now the chemists had not forgotten that for many centuries the early Greek notions of the structure of matter had prevailed, these notions being that all matter was made up of differing proportions of four fundamental "elements" — fire, air, water, and earth. Although this was no longer accepted, even in the 1700's, some of the force of this intellectual tradition remained. Here was the case of a kind of "air" reacting with ordinary air when heated, becoming fire and turning to water. From one "element" via a second "element" to a third "element."

The French chemist Antoine-Laurent Lavoisier emphasized this startling property of "inflammable air" some years after its discovery by giving it a name that reflected the transmutation. He called it "hydrogène" (converted in English to *hydrogen*) from the Greek word "hydor," meaning "water" and a Greek suffix "-genes," meaning "born" or "produced." Hydrogen, therefore, is something from which water is produced.

The Germans, who have less of a tendency for turning to Greek and Latin for their scientific words than do the French and British, named the new "air" in straight German. They, too, however, paid honor to the strange transmutation and called it "Wasserstoff," meaning "water substance."

Nowadays, of course, hydrogen has gained a fearful new importance through another transmutation it undergoes; not into a substance incorrectly labelled an element by the Greeks, but into a substance, helium, that is truly an element. This new transmutation provides the energy for the destructive force of the hydrogen bomb (see THERMONUCLEAR REACTION).

Hydrophobia

All human beings know fear. Some fears (of lions or homicidal madmen when these are in your vicinity) are normal and even sensible. However, there are abnormal fears, too, or *morbid* fears (from the Latin "morbidus," "sickly," the noun form of which is "morbus," meaning "disease"). Psychologists call such a morbid fear a *phobia*, from the Greek "phobos" (fear).

Phobias are subdivided according to subject. A familiar example is *claustrophobia*, which is a morbid fear of enclosed places (from the Latin "claustrum," meaning "a closed place"). Another is *agoraphobia*, the morbid fear of open spaces (from the Greek "agora," meaning "a market place," which, of course, was the chief open space in a Greek city). Thus, a claustrophobe may absolutely refuse to stay in a room with closed doors, while an agoraphobe may absolutely refuse to stay in a room with open doors, even when no conceivable danger is involved either way. There are dozens of specialized fears named in this way, even *panphobia*, which is a morbid fear of everything (from the Greek "pan," meaning "all") and *phobophobia*, which is a morbid fear of being afraid.

Less extreme cases (and more political than psychological) are extreme dislikes of things English or Russian, let us say. This is called *Anglophobia* or *Russophobia*, respectively.

But one phobia is actually a physical disease and not a state of mind. There is a virus infection that attacks the nervous system. A person with that disease cannot swallow. The attempt to swallow water or even just the sight or sound of water (which automatically makes him try to swallow) throws him into convulsions. The ancient Greeks considered these convulsions to result from a morbid fear of water and they called the disease *hydrophobia* from "hydor" (water).

This disease is generally transmitted through the bite of an infected animal, and in animals, it is usually called by its Latin name of *rabies* from "rabere" (to rave), since the animal is in such distress and agony that it behaves as though in a senseless rage. We usually call a dog with rabies a "mad dog."

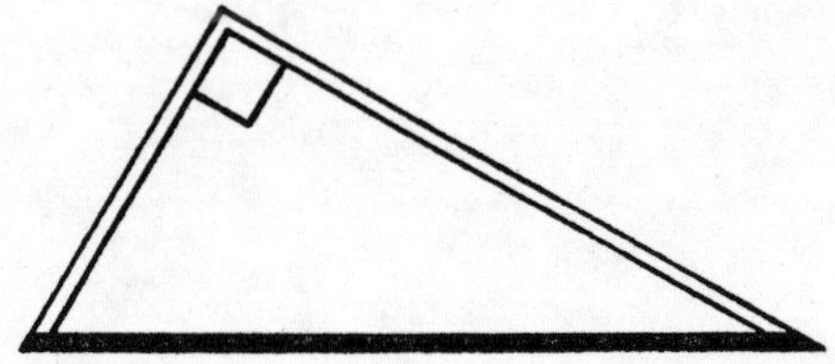

Hypotenuse

A LINE drawn on a perfectly level surface is *horizontal*. That is, it points toward the horizon at either end and not above or below it. *Horizon* itself comes from the Greek "horizo," meaning "bound," since the horizon bounds the visible earth in all directions.

A weight suspended over the surface would hang by a *vertical* line — one, that is, that points to the vertex, the topmost point of the sky. *Vertex* comes from the Latin "vertere" (to turn) and was originally used for the center of a whirlpool about which all turned, and was then applied to that point on the scalp about which the hairs seemed to turn when each lay in its natural position. Since this latter point is at the top of the head, more or less, *vertex* came to mean the top of anything.

A vertical line is *perpendicular* to a horizontal line, from the Latin "per" (through) and "pendere" (to hang); the vertical line hangs down and through the horizontal one as in the familiar plus sign (+). The angles formed in the plus mark are *right angles*. Our *right* comes from the Latin "rectus" which means several other things, including "upright." The right angle is formed when one of the lines stands exactly upright.

Of course, if the plus sign is tilted to form the multiplication sign (×) so that neither line is upright, the two are still perpendicular and the angles are still right angles.

The two lines of an angle meet at a point called the vertex (see above) because that point can be drawn topmost. If a right angle is so drawn (⌃) a third line drawn underneath will form a *triangle* (from the Latin "tres" meaning "three," so that a *triangle* is a figure with "three angles"). A triangle which includes a right angle as one of the three is called, specifically, a *right triangle*.

The third line "stretched under" a right angle to form the right triangle is called the *hypotenuse*, from the Greek "hypoteinousa" — "hypo" (under) and "teinein" (to stretch). What could be clearer?

Idiot

It is perhaps only human that there are a great many different words used to express mental deficiency, most of them slang; the vocabulary of insult is always great. Psychologists, however, have tried to make objective use of three of them to indicate various grades of mental deficiency.

A *moron* is only mildly deficient. He is capable of doing useful work under supervision. The term was adopted in 1910 by psychologists, and is derived from the Greek "moros" (stupid).

More seriously retarded is an *imbecile*, who cannot be trusted to do useful work even under supervision, but is capable of connected speech. Whereas *moron* has always applied to mental deficiency, *imbecile* referred originally to physical deficiency since the word is derived from the Latin "in-" (not) and "baculum" (staff); that is, it refers to a person too weak to get along without a staff. In the modern meaning, it is the mind that cannot get along without help.

Most seriously retarded is the *idiot*, one who is not capable of connected speech or of guarding himself against the ordinary dangers of life. This word has the oddest history of the three. The ancient Greeks were the most political of people. Concerning oneself with public business was the pet hobby of everyone. The Greek word "idios" means "private" so a Greek who, despite all this, was odd enough to concern himself only with his private business rather than with public business was an "idiotes." The Greek view concerning such a person is obvious since "idiotes" and "idiot" are the same word.

Of the more colloquial words, *fool* comes from the Latin "follis," meaning "bellows," obviously implying that a fool is someone whose words, though many and loud, are so much empty air. The slang expression "windbag" is the exact equivalent. *Stupid* is from the Latin "stupere" (to be stupefied; to be rendered speechless). Here the implication is of someone without words. Apparently, for one to be intelligent, his words must be neither too few nor too many, and in my opinion that's not a bad way of putting it.

Igneous

THE INTERIOR of the earth is at a high temperature, high enough to melt iron in the central depths. This heat arises from radioactive decay of certain elements within the earth (chiefly uranium, thorium, and one isotope of potassium).

The rocky crust of the earth is exposed to more heat with increasing depth and also to more pressure from the weight of the rock above. Eventually, heat and pressure combine to turn rock (which we know as hard and brittle) into a soft, doughy mass. Such rock is called *magma*, a Greek word meaning "dough," which is akin to "massein" (to knead).

This magma may work its way up toward the surface and, as the temperature and pressure both decrease, it slowly solidifies and becomes hard. Such rock may eventually be exposed on the surface of the earth, and the large crystals that make it up show that it has solidified slowly out of a liquid state. These rocks are called *plutonic* rocks, from the Greek god of the underworld, Pluto.

Occasionally, pockets of magma occur quite close to the surface, perhaps because of local accumulation of radioactive materials and this may break out as flows of molten rock (see VOLCANO), which quickly congeal and harden into small-crystal solids. These are *volcanic* rocks.

Together, plutonic and volcanic rocks are *igneous* rocks, since both are formed from the "fires" beneath the earth and "ignis" is the Latin word for "fire."

Another kind of rock is formed out of material worn away from solid land by the action of rain and rivers and carried down to the ocean where it settles out and sits at the bottom. Such settled material is *sediment* (from the Latin "sedimentum" which comes from "sedere," meaning "to sit"). As more sediment accumulates, the lower layers are compacted together by the pressure of the upper layers. The result is *sedimentary rock* and appears on the surface in regions where shallow seas once existed and are now gone.

Incisor

THE HUMAN mouth contains thirty-two teeth of different shapes, each shape designed for a special purpose. The eight teeth in front (four upper and four lower) are broad, and taper into a chisel-like cutting edge. These are the *incisors*, from the Latin "in-" (into) and "caedere" (to cut); they "cut into" food. Mammals such as rats, squirrels, and beavers have very prominent incisors which are constantly used for gnawing. These animals are examples of *rodents*, from the Latin "rodere" (to gnaw).

Next to the incisors are a total of four conical teeth, one on each side, upper and lower. They tear at food, rather than cut, and are particularly useful to meat-eating animals. They are prominent in dogs, for instance, and are sometimes called the "dog teeth" for that reason, though more often *canines*, which is much the same, since it comes from the Latin "caninus" (of the dog) which in turn comes from "canis" (dog).

The twelve rear teeth, three on each side, upper and lower, have flat irregular surfaces, between which food can be ground and which are sometimes called the *grinders* for that reason. Since just such a grinding action takes place in a mill, where grain is ground to flour between two millstones, these hind teeth are also called *molars*, from the Latin "mola" (millstone).

Between the molars and the canines are eight teeth (two on each side, upper and lower) that possess two conical points which are therefore called *bicuspids*, from the Latin "bi-" (two) and "cuspis" (point). They are also called *premolars*, because they come before the molars, the Latin "pre-" meaning "before."

The Latin word for "tooth" is "dens" (genitive, "dentis"), hence *dentist*, for a man who specializes in the treatment of teeth. The Greek word for "tooth," however, is "odous" (genitive, "odontis") and that gives the name to several specialized varieties of dentists. For instance, an *exodontist* is a dentist who specializes in tooth extraction. The Greek "ex-" means "from"; he takes "teeth from" a jaw. Again, an *orthodontist* is one who specializes in straightening teeth, from the Greek "orthos" (straight).

Indigo

Before 1856, the only dyes available to mankind for use on textiles were those that could be found in nature, and of those, only three were really good. That is, only three were brilliant in color, would not fade in air and sunlight, would attach firmly to textiles, and would not wash out. Consequently, the three were immensely valuable products for centuries (until the chemist learned to improve on them with hundreds of synthetic dyes, and threw the natural dyes out of business).

One of the natural dyes was the red-purple substance obtained from a small Mediterranean shellfish. Its production was the specialty of the ancient Phoenician city of Tyre; even today the dye is called *Tyrian purple*. It was so expensive, its use was restricted to royalty.

A second dye (orange-red) was obtained from the root of a plant called madder and is called *alizarin* from the Arabic "al asarah" meaning "the juice." To the Arabs it was so valuable, there was no need apparently to specify which juice.

The third dye (blue-purple) was obtained from an Indian plant which in Sanskrit was called "nili." The Arabs called it "al nil" (the Arabic "al" is simply "the") which the Spaniards further changed to "anil." The name "anil" has led to certain important chemical names (see ANILINE), but it is not what English-speaking people call the plant.

To the Romans the plant was simply "indicum" (Indian) and that has come down to us, again through Spanish, as *indigo*. The word was transferred from the plant to the dye and from that to the color. Indigo also gave rise to a number of chemical names, including that of an element.

In 1863, two German physicists, Ferdinand Reich and Hieronymous T. Richter, discovered a new element which, when heated, showed a bright spectral line (see SPECTRUM) that was indigo in color. They therefore named the new element *indium*.

Inertia

THE ANCIENT Greeks, more than any people who ever lived, had a profound appreciation of talent, both physical and mental. A person without the ability to do something well, without the possession of some art, was not a complete man. Later civilizations have inherited a little of this feeling and it still shows up in our word *inert*, which comes from the Latin "in-" (not) and "ars" (art). A person who has "no art" merely vegetates; he lacks an essential spark of life. And so the word is applied to anything without life, anything that is sluggish, heavy, unresponsive, resistant to change, and so on. (Of course, art also has an evil meaning, when it is used in the sense of wile rather than talent. A person without wile is *artless* and that kind of "no art" is considered a virtue.)

In 1687, the English mathematician Isaac Newton presented the world with three simple Laws of Motion on which all modern mechanics is based. The first law is this: "Every body persists in a state of rest or of uniform motion in a straight line unless compelled by external force to change that state."

This means that a brick resting on a board would rest there through all eternity unless it were pushed or pulled and made to move. Left to itself, it would never move. This seems to stress the inertness of matter; it raises inertness to the state of a natural law. For this reason, Newton's First Law is called the principle of *inertia*.

Of course, the law also states that if the brick were hurtling through space, it would continue moving forever, unless stopped. That doesn't seem very "inert" of it, but actually it is. Remember that *inert* implies a resistance to change. You could look at it this way: that the brick having started moving, at last, is now too "lazy" to stop moving, or even to change its direction of motion, unless made to do so by some external force (see FRICTION).

That, after all, is still inertia, but in another guise.

Infinite

THE ORDINARY objects that surround us have a beginning and an end. A piece of paper, a ruler, a locomotive, the United States of America — all have some end on all sides such that if you go past that end, you reach a place where the piece of paper or the ruler or the locomotive or the United States of America does not exist. The universe itself, according to Einstein's theories, has an end. The Latin word "finis" means "end" (we see the Latin word on movie screens sometimes, for that reason) so objects that come to an end are *finite*.

Despite the finiteness of the universe, not everything comes to an end. If you were to start counting and continue counting until you came to an end, you would find that there was no end. After every number, no matter how large, it is always possible to add another number. The total number of numbers, in other words, is not finite. It is (using the Latin prefix "in-" which means "not") *infinite*.

The quality of being endless is *infinity*. You may speak, for instance, of the infinity of numbers or the infinity of ideas. Infinity is not, however, itself a number. You must not imagine infinity to be the "last number," since there is no last number. Mathematicians find it useful to use the symbol (∞) to represent a condition of endlessness, but the symbol is usually read by everyone (even mathematicians, themselves, when in a hurry) as *infinity* which, strictly speaking, is wrong.

It is also possible to imagine something that is "infinitely small." In the number system, for instance, one is the first number and as close as you can get to zero without actually being zero. However, if we consider fractions, one-half is smaller than one, one-quarter still smaller, and one-eighth smaller still. In fact, no matter how close to zero the fraction gets, one can get still closer without actually reaching zero by increasing the denominator without increasing the numerator of the fraction. An infinitely small thing is said to be *infinitesimal*.

Influenza

THE HEAVENLY bodies were thought in ancient times to govern the fates of men, and the study of the exact manner in which they did this was known as *astrology*, from the Greek "astron" (star) and "logos" (word). An astrologer talks about the stars, in other words. Usually, the suffix "-logy" implies a branch of respectable science, but the false notions of astrology have so discredited the word that the science of the stars is today known as *astronomy*, instead. The suffix "-nomy" comes from the Greek "nemein" meaning "to arrange." Thus an astronomer was one who studied the arrangement of the stars, originally, but the word has come to mean one who studies everything about the stars.

The mysterious power residing in the stars was supposed by the astrologers to flow down to earth and into men, governing them. The Latin word for "flow in" is "influere" so the stars were said to have influence over men.

In the days before the real cause of disease was understood, it was natural to assume that sickness, like everything else, was caused by the influence of the stars, so one of the most common diseases was called just that — *influenza* (which is Italian for "influence").

Another disease which reveals prescientific notions of cause is *malaria*, which is now known to be caused by a one-celled animal that infests our red blood cells and is spread from person to person by mosquitoes. In Italy, during the decline of the Roman Empire and during the worse times of the barbarian invasions that followed, fields were abandoned and allowed to revert to swampland along the coast. Mosquitoes flourished among the swamps, and malaria spread.

The Italians noticed the connection between malaria and the swamps, but overlooked the mosquito. To them, the mosquito was only a nuisance; the real fault lay with the bad air that hovered over the swamps. (It was undoubtedly musty with decaying vegetation, and smelled.) The Italian words for "bad air" are *mala aria* and hence the name of the disease.

Infra Red

LIGHT BEHAVES as waves in at least some ways. The waves of a particular ray of light will cover a fixed distance with each undulation. This distance is the *wave length* of that light. Such wave lengths are generally measured in *Ångstrom units* (abbreviated, Å), one of which is equal to a ten-billionth of a meter. (It is named after the Swedish physicist Anders Jönas Ångstrom, one of the pioneers of spectroscopy; see SPECTRUM.)

Light with wave lengths of about 7000 Å seems red to our eyes. A wave length of 6500 Å strikes us as orange, 5900 Å as yellow, 5400 Å as green, 4800 Å as blue, and 4200 Å as violet. The spectrum gives the entire range of visible colors from 7200 to 4000 Å.

But visible light is not the only form of radiation. A photographic plate placed beyond the violet end of the spectrum, where nothing is visible, is strongly fogged, so there is some sort of radiation there. This is sometimes called *black light*, but its more formal name is *ultraviolet* light, "ultra" being a Latin word meaning "beyond." The wave lengths of ultraviolet light range from 4000 Å down to 100 Å. (X rays and gamma rays have still shorter wave lengths.)

Similarly, there can be light beyond the red at the opposite end of the spectrum. The astronomer William Herschel (see URANIUM) noticed, in 1800, that a thermometer placed beyond the red showed a rise in temperature. There are radiations there, too. These are sometimes called *heat rays* because of this heating effect. The more formal name, however, is *infra red* light, "infra" being a Latin word meaning "below." Why "below" rather than "above"? Because red light contains less energy than any other form of visible light. In terms of energy content it is at the bottom of the scale. Infra red, with still less energy, is below it.

Infra red light has wave lengths of from 7200 Å all the way up to 3,000,000 Å. Beyond it lie the radio waves with still longer wave lengths.

Insect

INSECTS ARE the great pioneers of the animal kingdom. They were the first animals to invade the land, the first to learn to fly, the first to develop complex social groups. The pioneering has paid off; there are more different kinds of insects in existence today than the total of all kinds of all other animals put together.

One of the distinguishing characteristics of insects is the division of their bodies into three relatively thick regions — head, thorax and abdomen — which are connected by short and sometimes very thin stalks. (We have the expression "wasp waist" for an unusually small waistline.)

The insect outline is cut into, so to speak, between the main regions. The Latin word for "to cut into" is "insecare" (past participle, "insectum"), hence, *insect.*

Many insects emerge from the egg in the form of wormlike creatures that bear no resemblance to the adult insect that laid the egg. Out of a butterfly's egg, for instance, comes a caterpillar. This immature form is a *larva* which is the Latin word for "ghost." The connection is rather farfetched, but just as a ghost originates from a human being but is quite different in appearance, so a larva originates from an ordinary insect but is quite different in appearance.

Eventually, the larva enters a quiet stage during which its body is reorganized into the adult form. It sometimes weaves a silken covering, a *cocoon* (from the French "cocon," meaning "a little shell") as protection in the meantime. In this quiet, reorganizing state, the insect is a *pupa,* a Latin word meaning "doll" or "puppet," since in that stage it shows no visible life. The pupa stage of some butterflies is golden in color so another term for pupa is *chrysalis* from the Greek "chrysos" (gold).

An insect in the pupa stage or in the late larval stage is also sometimes called a *nymph.* The nymphs in Greek mythology were semigoddesses who were ever young and ever beautiful. The Greeks gallantly applied the term to young girls on the point of marriage and adulthood and we spoil it by applying it to insects on the point of adulthood.

Insulin

THERE ARE glands which, unlike the liver (see BILE), do not pour the chemicals they manufacture through a duct, but transfer them directly into the blood stream. These are the *ductless glands*, or the *endocrine glands*. The latter name comes from the Greek "endon" (within) and "krinein" (to separate); the fluid is separated within the body and distributed directly into the blood. Such glands produce hormones (see HORMONE) and it is for this reason that doctors who specialize in the workings of hormones are called *endocrinologists*.

The pancreas is a gland with a duct, and a pancreatic juice travels through the duct into the intestine. In 1869, however, a German pathologist, Paul Langerhans, discovered little clumps of cells scattered throughout the pancreas that were different from the rest. In his honor, these are called the *islands of Langerhans* (perhaps the most romantic-sounding name in the body).

In 1889, it was discovered that a dog from which the pancreas had been removed did not live more than a few weeks and for those few weeks he showed symptons similar to those of a human disease called *diabetes mellitus*. *Diabetes* is a Greek word meaning "something which goes through." It is applied to those diseases which involve an overproduction of urine; liquid just seems to "go through" the body without pausing. In diabetes mellitus, sugar is not handled properly by the body; it accumulates in the blood and spills over into the urine. The resulting sweetness of the urine is described by *mellitus* which comes from the Latin "mel," meaning "honey."

In 1916, the British physiologist Sharpey-Schafer suggested that the islands of Langerhans were a group of ductless glands buried in an ordinary gland and that they produced a hormone that controlled the manner in which the body handled its sugar supply. Since the hormone was produced by "islands," he suggested the name *insulin* from the Latin "insula" (island). The hormone was indeed eventually isolated and Schafer's name stuck.

Iodine

It is common to have a liquid turn into a vapor when heated. It is less common to have a solid substance turn into a vapor directly without ever going through the liquid state. The best-known example these days is solid carbon dioxide, which has the appearance of a cloudy ice, but is much colder, and which doesn't turn to liquid when warmed but to gas. Because of this absence of liquid and wetness, the common name for solid carbon dioxide is Dry Ice.

A number of other substances behave the same way. A good example is iodine. We are most familiar with iodine when it is in solution in a mixture of water and alcohol. This is *tincture* of iodine, from the Latin "tinctura" (dyeing), from "tingere" (to dye), which is the source, also, of the common word *tint.* In pharmacy, alcoholic solutions are commonly called tinctures because many dyes will dissolve in alcohol but not in water. Tincture of iodine is reddish brown in color.

Iodine itself, however, is an element which is solid at room temperature and forms slate-gray crystals. If some is heated gently in a test tube, it will not liquefy but will form a beautiful violet vapor which will solidify into gray crystals again in the upper portion of the test tube where the glass is cooler. The iodine has been raised to a higher place in this manner.

The Latin word for "high" is "sublimis" from "sub-," meaning "under," and "limen," meaning the lintel (upper wooden crossbar) of a door. Something just under the top crossbar of a door is relatively high. So lofty ideas or thought or behavior are said to be "sublime," and when a solid turns directly into vapor it, too, is said to *sublime.*

The violet vapors of iodine were first observed in 1811 by the French chemist Bernard Courtois. He was working with seaweed ash and at one stage usually added sulfuric acid. He might have added too much once, for a violet vapor turning into gray crystals was formed, and Courtois had discovered a new element. He named it *iodine* from the Greek "iodes" (like the violet) and commemorated the first sight of those vapors in that way.

Ion

IN ELECTROLYSIS, parts of a molecule seem to travel to one electrode and parts to the other. The water molecule, for instance, breaks up into hydrogen and oxygen, the hydrogen appearing about the negative electrode, or cathode, and the oxygen appearing about the positive electrode or anode (see ELECTROLYSIS).

The traveling parts of the molecules were given the name *ion* in 1830 or thereabouts by the British physicist Michael Faraday, from a Greek word "ienai" meaning "to go." The Greek word "ion" means "going." After all, they were going to one electrode or the other. The ions that traveled to the cathode were *cations;* those that traveled to the anode were *anions* (both words are pronounced in three syllables). Exactly what ions were, however, remained a mystery for about fifty years.

Then, in 1884, a 25-year-old Swedish physical chemist, Svante August Arrhenius, presented a dissertation at the University of Uppsala, with which he hoped to earn the degree of Doctor of Philosophy. He suggested that under the influence of the electric current, molecules broke up to form atoms or groups of atoms that carried electric charges. Those with a negative charge were anions, attracted to the positive electrode, while those with a positive charge were cations, attracted to the negative electrode.

Since chemists had, at that time, never heard of atoms carrying an electric charge, his theory was considered quite ridiculous and he got his degree with a minimum passing grade. However, in 1903, Arrhenius received the Nobel Prize in chemistry for that same dissertation.

Between 1884 and 1903, you see, the negatively charged electron had been definitely shown to exist and to form a part of all atoms (see ELECTRON). It was recognized that an atom or group of atoms might lose one or more electrons to become positively charged, or gain one or more to become negatively charged. In the former case, a cation resulted; in the latter, an anion.

Irrational Number

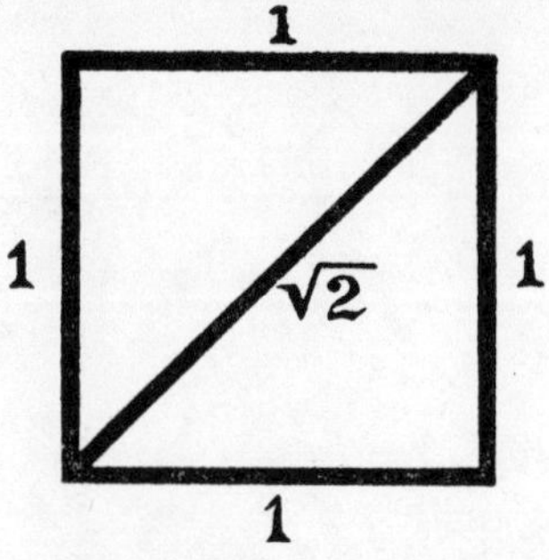

ANY NUMBER which can be expressed as the ratio of two digits is a rational number (see FRACTION). To a nonmathematician, who doesn't think of *rational* as being derived from *ratio*, there is a natural tendency to think of such numbers as being "reasonable" or "sensible" ones, because the word *rational* means that in the ordinary vocabulary. It comes from the Latin "rationalis" which in turn comes from the verb "reari," meaning "to think." So does the word *ratio*, which accounts for the confusion.

When fractions were first discovered it seemed logical to assume that any conceivable number could be written as a fraction. For instance, a number halfway between $\frac{1}{2}$ and $\frac{1}{3}$ is $\frac{5}{12}$. A number halfway between $\frac{5}{12}$ and $\frac{1}{3}$ is $\frac{9}{24}$. To express some numbers it might be necessary to use big digits as, for instance, $\frac{28067048}{57134097}$, but that doesn't affect the principle of the matter.

But now what about square roots? If you restrict yourself to digits only, the square root of 4 is 2 and that of 9 is 3 (see SQUARE ROOT). But there is no digit that is the square root of 8, for instance. Yet what if you use fractions? If you multiply $\frac{14}{5}$ by $\frac{14}{5}$, you have 7.84. That makes $\frac{14}{5}$ almost the square root of 8, but not quite. Try a slightly larger fraction and work out $\frac{141}{50} \times \frac{141}{50}$. This comes to 7.9524, which is better. Multiply $\frac{707}{250} \times \frac{707}{250}$ and you have 7.997584, which is still better. It certainly appears that if you keep adjusting fractions, you will eventually find one that will, when multiplied by itself, give you exactly 8.

But this is not so. You can get awfully close to 8 but you can never hit it exactly. There is no fraction which is the square root of 8 (or of the vast majority of other digits). The square root of 8 (and of most other digits) simply cannot be expressed by a ratio of two digits in any way. The square root of 8 (and most other digits) is an *irrational number*. To the ancient Greeks who discovered this (and to many a schoolboy since), this has seemed so odd that irrational numbers seemed irrational indeed, in the ordinary sense.

Isomer

UNTIL 1800, chemists were quite certain that every different compound was made up of a different combination of elements (regardless of what they thought the elements were); or if the same elements were found in different compounds, at least they would exist in different proportions. And this is generally true in the realm of inorganic chemicals.

Organic chemicals (see ORGANISM) are, however, made up for the most part of only a handful of elements. Carbon, hydrogen, oxygen, and nitrogen occur most frequently. And since there are uncounted thousands of organic compounds, it is not surprising that occasionally two should be found with molecules made up of equal numbers of the same atoms. By 1830, enough such pairs had been found, so that the Swedish chemist Jöns Jakob Berzelius thought it wise to propose a name. He suggested that compounds of identical composition but different properties be called *isomers*, from the Greek "isos" (equal) and "meros" (part); they were made up of "equal parts" of the various elements.

For a number of years, the reason for the existence of isomers was not quite understood. Then, in 1874, the French chemist Joseph A. Le Bel and the Dutch chemist Jacobus H. Van't Hoff independently suggested that the carbon atom attached itself to four other atoms arranged in specific positions about itself. Obviously then, chemicals might have molecules made up of the same number of the same atoms, but in different arrangements.

As a result of this, it sometimes happens that when a compound is discovered which differs from an old established compound only in atom arrangement it is named by adding the prefix "iso-" to the older name, as a short version of "isomer of." For instance, in 1818, one of the amino acids (now known to be found in proteins) was first isolated and was named *leucine* from the Greek "leukos" (white) because it showed up as white crystals. (This is a poor excuse for naming, by the way. The majority of organic chemicals are white.) When, in 1905, another amino acid was discovered, differing from leucine only in a minor detail of atom arrangement, it was named ***isoleucine***.

Isosceles

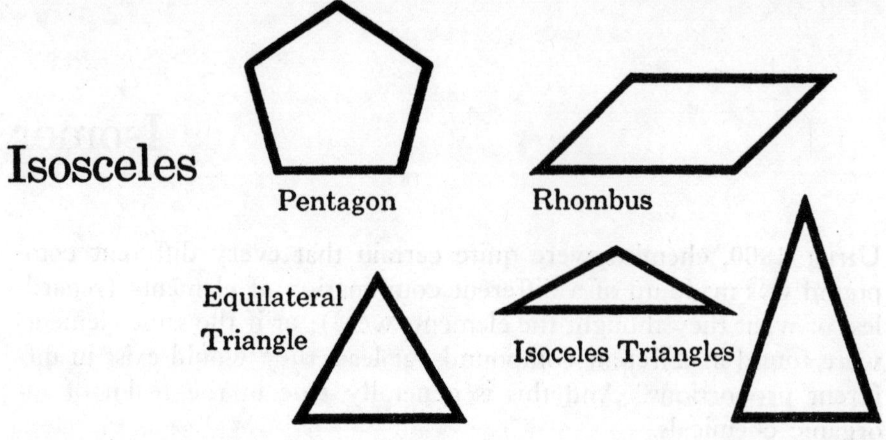

A CLOSED FIGURE made up of straight lines may contain any number of sides from three up, and the same number of angles, if the lines don't cross. The Greek word for "angle" is "gonia" and for "much" is "polys." A figure containing a number of angles is, therefore, a *polygon.*

The particular variety of polygon is named from the Greek numbers. A five-angled figure is a *pentagon* ("pente" is "five"); a six-angled one is a *hexagon* ("hex" is "six"); an eight-angled one is an *octagon* ("okto" is "eight") and so on. On this system, a four-angled figure ought to be a *tetragon* and a three-angled one a *trigon.* These last two can, indeed, be found in the dictionary but they are practically never used.

Instead, the Latin equivalents are used. A four-angled figure is a *quadrangle* (quattuor" being Latin for "four") or, even more commonly, a *quadrilateral* ("four sides," see PARALLEL). The three-angled figure is a *triangle* ("tres" being Latin for "three").

A polygon with all sides equal is *equilateral*, from the Latin "aequus" (equal) and "latus" (side). The only equilateral polygon with a special name of its own is the equilateral quadrangle. This is a square (see PARALLEL) if all the angles are right angles, or a rhombus if they are not. (The word *rhombus* comes from the Greek name for a small block of wood of that shape which was placed on a string and whirled as a noisemaker during certain festivals. The Greek word "rhembein" means "to whirl.")

A triangle with all sides equal is an *equilateral triangle.* One with only two sides equal is an *isosceles triangle;* the word *isosceles* comes from the Greek "isos" (equal) and "skelos" (leg). (A man's two legs, standing apart, make an isosceles triangle with the ground.)

Isotope

UNTIL THE middle 1800's there seemed no orderliness in the nature of the various elements that were being discovered. In 1869, however, the Russian chemist Dmitri Ivanovich Mendeléev listed the then-known elements according to the weights of their atoms. He showed that similar properties turned up in regular periods so that the listing was called the *periodic table*. By use of it, Mendeléev predicted the properties of still-unknown elements and lived to see those prophecies come true.

For thirty years, the periodic table met every test. Then, in 1896, uranium was found to give off strange forms of radiation. In doing so it broke down to another substance which broke down to still another and so on. The element thorium also behaved in this fashion.

Chemists located a series of breakdown products, over 40 of them, all told, each with its own properties, but there was simply no place in the periodic table to put these new elements. For instance, one substance was "radium G." It behaved like lead, chemically, but it gave off radiations whereas ordinary lead did not. It fitted nowhere in the table. Then, too, three different gases were discovered as products of these breakdowns, but there was only one place in the periodic table to put them.

Several scientists, notably the British chemist Frederick Soddy, solved this dilemma in 1913. They suggested that a particular element might have different types of atoms. It has since been definitely proven that this is so. All atoms of a given element have the same number of protons and are therefore alike in ordinary chemical ways. They may vary, however, in the number of neutrons and will behave differently in other ways, in whether they break down or not, their particular manner of breakdown, and so on.

These different atomic varieties of a single element belong in the same place in the periodic table and Soddy made that very point by calling such atomic varieties, *isotopes* from the Greek words "isos" (equal or same) and "topos" (place).

Ketone

In many organic compounds there occurs a carbon-oxygen combination which can, in turn, be attached to two other atoms. One of these other atoms is often hydrogen.

Compounds containing such a carbon-oxygen-hydrogen combination can be obtained by removing two hydrogen atoms from an alcohol molecule; i.e., by *dehydrogenating* it. In 1835, the German chemist Justus von Liebig recognized this fact and suggested the name *aldehyde* for such compounds as an abbreviation for "*al*cohol *dehyd*rogenated." The name was accepted and the official suffix for the names of such compounds is "-al" now.

The carbon-oxygen combination need not be attached to a hydrogen atom, however, but to two carbon atoms. Again, it was von Liebig (in 1831) who worked out the actual molecular structure of the simplest of the compounds containing such a combination. He suggested the name which, in English, is *acetone.* The "acet-" part comes from acetic acid (see ACID), from which acetone can be formed. The "-one" ending was originally used to indicate a chemical that was weaker than the one from which it was prepared, and certainly acetone is a milder substance than is acetic acid.

Now in English, *c* before *e* is pronounced "s" so that *acetone* is pronounced "asetone." But in Latin, *c* is always pronounced "k" and in German, the letter *c* is never used except in foreign words, anyway. Liebig's name for acetone was therefore "aketon" which is actually more correct.

In 1848, the German physiologist Leopoldus Gmelin wanted a word to express all compounds that contain the atom combination that was found in acetone. To obtain such a word, he merely dropped the *a* from "aketon" and used what was left. Consequently, our word for compounds of the acetone type is *ketone* to this day. Furthermore, the "-one" suffix in organic chemistry has lost its original meaning and is now used to indicate that the chemical in question is a ketone.

Lemur

PRIMITIVE PEOPLE, to whom many aspects of the physical world are strange and terrifying, because not understood, fill their environment with imaginary gods, demons, ghosts, and monsters. The Greeks and Romans were no exception and some of their words for ghosts are hidden in scientific terminology (see INSECT).

The ancient Romans, for instance, talked of horrible night-prowling spirits called "lemures." The word may come from older words meaning an open, gaping mouth, so you can imagine the kind of spirit it was.

Early explorers of Madagascar came across small animals that prowled the night, also. What's more, they were so timid and elusive, one caught only glimpses of them, as though they were ghosts. So, despite the fact there was nothing of the devouring horror about them, they were called *lemurs*.

The lemurs, which are primitive monkeys, exist mainly in Madagascar, but also in a few places in southeast Asia. Some romantic people imagine that there was once a continent in the Indian Ocean, connecting Madagascar and Malaya, and that most of it sank, marooning its lemur inhabitants at the two ends which remained above water. This imaginary continent, which was named *Lemuria*, has entered fiction as a kind of second Atlantis.

One type of lemur has the bone (the tarsus) of its foot very elongated so that it seems to sit on stilts. It is called the *tarsier*, for that reason. It is a little night-creature with enormous eyes (for its size) in the front of its head. It is completely harmless, but seeing those large eyes suddenly in the dark, staring solemnly, must give one quite a shock. So it is called the *spectral tarsier* or even the *spectral lemur*. Since a specter (from the Latin "spectrum," meaning "vision") is again some kind of ghost, a spectral lemur is obviously a ghosty ghost indeed.

Lepidoptera

THE INSECTS are the one form of invertebrate life that have developed wings. Insect wings are membranous films that differ from the wings developed by vertebrate creatures in not being converted fore limbs. It was the wings that apparently impressed those who classified the insects into separate groups, for almost all the main divisions have names that are derived in part from the Greek word "pteron" (feather or wing).

For instance, the most familiar insect, the housefly, differs from most insects in having two wings rather than four. It belongs to the order *Diptera*. The Greek prefix "di-" means "two" so you see the order includes the "two-winged" insects.

The beetles don't look as though they have wings at first glance, but they do. The hind wings are ordinary wings but are folded beneath the fore wings which have evolved into horny opaque sheaths that fit closely over the body and hide and protect the gauzy hind wings. Beetles belong to the order *Coleoptera*, from the Greek "koleon" (sheath), hence the "sheath-winged" insects.

The most spectacular wings of the insect world belong, of course, to the butterflies and moths. These are large wings covered with minute scales that come off as a dust when the wing is handled. The only possible name would seem to be *Lepidoptera*, from the Greek "lepis" (scale), or the "scale-winged" insects. This name was given them by Karl Linnaeus in 1735. He first classified the insects, and this was one of the names that was never changed.

The colors of the Lepidoptera wings are often startlingly beautiful and in the early 1940's the chemical constitution of the compounds giving rise to these colors was worked out. The compounds were found to contain a double ring of atoms made up of six carbons and four nitrogens. Compounds containing this particular ring system are called the *pteridines* after their source. The most important pteridine is a rather complicated one called *pteroylglutamic acid*. This is one of the B vitamins essential to all life, and a Greek scholar might wonder what "wing" was doing in the name.

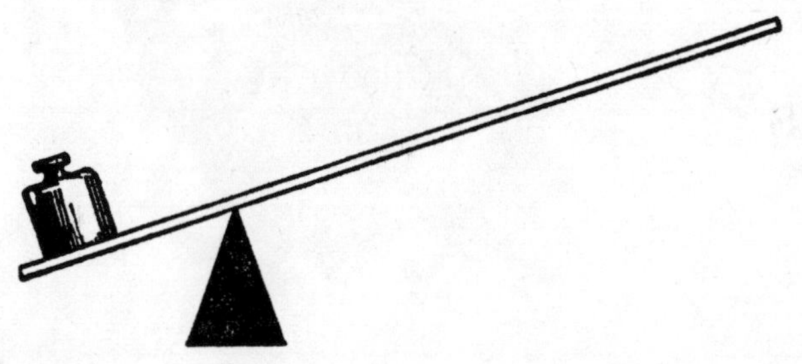

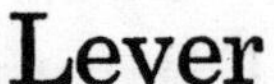

Lever

The lever is really a lopsided seesaw (see moment). Imagine a seesaw eleven feet long with a fulcrum placed one foot from one end. A force of one pound at the long end has as much of a turning effect on the seesaw as a force of ten pounds at the short end. The moments are equal since 1 (pound) × 10 (feet from fulcrum) is equal to 10 (pounds) × 1 (foot from fulcrum).

A 10-pound weight at the short end of this seesaw can therefore be lifted by pushing downward at the long end with a force of one pound. If the lever were long enough and lopsided enough, you could, with a one-pound force, lift 100 pounds or 10,000 pounds. In fact, the Greek mathematician, Archimedes, who discovered this principle of moments, said, "Give me a place to stand on and I can move the Earth."

Where does the extra force come from? At the expense of distance. The long end of the lever in the case described here moves downward ten times the distance that the short end moves upward. Force times distance is the same at both ends.

When you're prying up a huge boulder, after all, it may be that you need only pry it up one inch to send it hurtling down a hill at your enemy. So you sacrifice distance by forcing a long lever under one end of the boulder, propping the lever on a stone near the boulder as a fulcrum, then putting your weight on the long end, not minding if you have to push down three feet if only you can get that ton of rock to move up the crucial inch.

You might also sacrifice force for distance by using a lever wrong-way-round, as in a catapult. Then, by dropping a heavy weight at the short end of a lever and moving it down a short distance, a lighter stone at the long end will sweep up through a long arc and go sailing off against a fortress. It is because *levers* are so often used to lift something against gravity that they are called by that name, since the Latin word for "to lift" is "levare."

Liberal Arts

THE SUBJECTS taught in colleges are extremely varied. They include a large number of sciences, foreign languages, history, literature, economics, and so on. All such subjects are spoken of as *liberal arts*, as opposed to the technical or professional subjects studied in trade schools, law schools, medical schools, and others.

In ancient times, the liberal arts were exactly seven in number and were sometimes called, for that reason, the seven liberal arts. They were: arithmetic, geometry, astronomy, music, grammar, logic, and rhetoric. These were the "higher" arts and not everyone might have access to them. Slaves, for instance, might need a rudimentary education in order to be properly useful but the higher arts were for free men only. In Latin "free men" is "liberi" so these were the "liberal" arts.

The seven liberal arts were divided into two groups. The first four — arithmetic, geometry, astronomy, and music — made up the *quadrivium*, which, in Latin, means "a place where four roads meet," from "quattuor" (four) and "via" (road). In the mind of the student studying the quadrivium, four roads to learning met, so to speak.

The remaining arts — grammar, logic and rhetoric — were, naturally, the *trivium* ("a place where three roads meet," from "tres," meaning "three," and "via," "road"). This idea of a common place (hence, *commonplace*) is at the root of the word *trivial*. The trivium was considered less important, and the student in a hurry could omit them and study only the quadrivium. But it would be a mistake to consider the trivium trivial!

A student who mastered the seven liberal arts was, obviously, a Master of Arts, a degree still awarded by universities. *Master* comes from the Latin "magister," which in turn comes from "magnus" (great). A less thorough mastery entitled one to be a Bachelor of Arts, *bachelor* being a word which originally referred to a knight's squire, a man still young enough to have to follow another and who had not yet attained to being his own "master."

Libration

The moon travels once about the earth in 27.32166 days. It also rotates about its axis once every 27.32166 days. (This is not a coincidence but is a situation that is forced upon the moon by the earth's tidal drag.)

Since the period of rotation and of revolution are equal, the motion of the moon around earth and around its own axis is such that it always presents the same face to the earth. Or at least it would do so, if both motions were exactly even.

The moon's rotation is quite steady and even, to be sure, but its velocity of revolution about the earth varies with the distance between them. At one point in its orbit, it is as close as 221,000 miles to the earth and at another point as far as 252,000. Between those points it is at intermediate distances.

When the moon is relatively close to the earth, it moves more quickly than usual in its orbit and revolves a little too quickly for its period of rotation. The result is that the moon seems to turn a little from east to west so that we can see a bit about its eastern edge. This is made up for in the other half of the moon's orbit where it is relatively distant from earth and moves more slowly than usual so that the moon turns a bit from west to east (apparently) and we can see a small way about the western edge.

To a person watching the moon night after night, then, it seems to oscillate slowly about its axis, first to the east for two weeks then to the west, repeating over and over again. It behaves something like the scales of a balance, slowly oscillating as they come to rest (except, of course, that the moon is not coming to rest but continues to oscillate indefinitely). The Latin word for a balance is "libra" and this apparent oscillation of the moon about its axis is called *libration* (east-west).

The moon's axis is slightly tilted with respect to the earth so that sometimes we can see over the northern edge and sometimes over the southern. This is libration, too, of the north-south variety. If we count the maximum librations of both types, we can, at one time or another, see 59 per cent of the moon's surface altogether.

Lichen

LICHENS ARE simple plants but in one respect are very unusual. They offer the best example in existence of two life forms in equal partnership. A lichen consists of two parts: an alga and a fungus (see PLANKTON). The alga, possessing chlorophyll, is capable of manufacturing food and tissue components by use of the energy of sunlight. The fungus, for its part, provides water, salts, and anchorage. Each is better off for the association, and such an association is called *symbiosis*, from the Greek "syn-" (together) and "bios" (life); a "life together," in other words. It was the German botanist Heinrich Anton de Bary who, in 1873, first suggested the word.

Lichens grow everywhere, spreading over soil, over the bark of trees, over leaves, even over naked rock. In fact, the observation of life in the form of lichen working its stubborn way up a rock, like a fire licking upward along a tree trunk, may be what gave *lichen* its name, since the Greek "leichein" means "to lick," and their word for "moss" is "leichen."

Lichens may be divided into two types depending on the nature of the algal component. The alga is sometimes of the blue-green type and then the lichen is a *phycolichen*, from the Greek "phykos" which means "seaweed" and is therefore equivalent to "alga" which, after all, is just Latin for "seaweed." More often, the alga is of the bright green type and then the lichen is an *archilichen*, from the Greek "archaios," meaning "from the beginning," since these lichens are the more primitive.

There is one highly dramatic possibility in connection with lichens. There are on the surface of the planet Mars green patches which advance and retreat with seasons. In the northern summer, when the northern icecap melts, green patches advance in the northern hemisphere and retreat in the southern. Half a Martian year later the situation is reversed. The green color resembles that of some lichens and lichens, moreover, are the one form of life that may be sturdy enough to survive the severe Martian conditions. Perhaps when we land on Mars, someday, we'll find that lichens (or something like them) are licking along the Martian surface, too.

Liver

All glands produce and liberate some chemical or fluid. Although the word *gland* (from the Latin "glans," meaning "acorn") was originally applied to small organs, some glands are quite large.

The largest is the liver, which weighs nearly four pounds in man. It produces a juice that is delivered through a duct into the intestines. The word *liver* comes from the Anglo-Saxon and may, just possibly, have some connection with the word *life*. (In German, the word for "liver" is *Leber* and for life is *Leben.*) Certainly, the ancients, probably impressed by the sheer size of the organ, thought it the very center of man's life and health.

We ourselves, for instance, usually talk as though we think the heart is the center of life and the seat of emotions. We are "brokenhearted" or "fainthearted"; we "take heart" or "lose heart." Well, the ancients thought the liver was the seat of the emotions and used similar expressions with "liver" in them. Few of those old expressions survive now, but we still sometimes speak of cowards as being "lily-livered." Their liver, in other words, is pale with fear and so they, themselves, are afraid.

The second largest gland in the body is the *pancreas*, which is less than a tenth the size of the liver. Its name comes from the Greek "pan" (all) and "kreas" (flesh) because it is all flesh (that is, meat) without bone or fat. It is located just under the stomach and when the pancreas of a calf or other animal is bought as an item of food at the butcher's, it is called *stomach sweetbread*. The word *sweet* is a corruption of an Anglo-Saxon word for glandular food in general, while *bread* is used here in its old general meaning of "food" as in "Give us this day our daily bread."

Another gland, the *thymus*, which occurs in the neck region of young animals (and which supposedly gets its name from a fancied resemblance to a bunch of thyme — a plant that the Greeks called "thymon") is called *neck sweetbread* at the butcher's.

Lymph

The very smallest blood vessels have walls so thin that the watery portion of the blood can easily leak out. It does exactly this, bathing the cells and forming the *interstitial fluid* of the body. (It is called this because it fills the interstices between the cells. The word *interstice* comes from the Latin "inter" (between) and "sistere" (to stand); an interstice is something that "stands between.")

The interstitial fluid has not left the blood vessels permanently. It drains off into tiny vessels that join into larger ones and finally into two main ones that travel up the chest to the neck where they join veins and it is there the interstitial fluid rejoins the blood stream. The larger of the two main vessels is on the left and is called the *thoracic duct*, from the Latin "thorax" (chest), which in turn comes from the Greek "thorax" (a corselet), and "ducere" (to lead).

The interstitial fluid is like the blood in most ways. The most conspicuous difference arises from the fact that the cells in blood which give it its color cannot pass through the small vessels. The interstitial fluid is, therefore, colorless and this is reflected in its name, for it is called *lymph*, from the Latin "lympha" (clear water) — whence we get the word *limpid*. The vessels which contain interstitial fluid are the *lymph vessels*.

Here and there along the course of the lymph vessels (particularly on the side of the neck, under the jaw, in the armpits, elbows and groin) are small swellings. The early anatomists thought the swellings resembled acorns in shape and since the Latin word for "acorn" is "glans," the swellings were called *lymph glands*.

The lymph glands form cells that are called *lymphocytes*. (The suffix "-cyte" or the prefix "cyto-," when used in biological terms, mean "cell." They come from the Greek "kytos," meaning "a hollow place," so we have here the same sort of poor derivation from which the word *cell* itself suffers. See PROTOPLASM.) Lymphocytes are primarily bacteria-fighters and during an infection, lymph glands become more active. They swell and become painful, and mothers these days keep a sharp eye out for "swollen glands" whenever children grow feverish.

Magnet

Thousands of years ago, men were amazed at finding pieces of black mineral that attracted iron. It is said that the early Greek thinker Thales of Miletus (a city in Asia Minor) first studied the phenomenon. He got samples of the material from Magnesia (another city in Asia Minor) so he called the mineral magnes. This has come down to us in the word *magnet*, while the mineral is today called *magnetite*.

The Roman naturalist Gaius Plinius Secundus (usually called Pliny the Elder) confused the magnes of Thales with another black mineral, which he also called magnes. In medieval times, Pliny's books were copied by hand, sometimes by people who were not very careful, or perhaps not very learned. Distortions crept in. Pliny's mistaken "magnes" was mangled further and misspelled as "manganese."

In later times, Pliny's magnes was used in glassmaking to wash out the greenish color that resulted from iron impurities in the raw materials. The mineral, therefore, was named *pyrolusite* from the Greek words "pyr" (fire) and "louein" (to wash). (The "fire" referred to the strong heating required to make glass.)

In 1774, the Swedish mineralogist Johan Gottlieb Gahn isolated a new metal from pyrolusite and, in naming it, went back to the medieval, mistaken, and misspelled name of *manganese*. That name has been accepted ever since.

In ancient times, another mineral (a white one) had been discovered in the region of Magnesia. (This may not necessarily have been the same city as the first; there were three cities so named in ancient Greek territories.) The Romans called this mineral magnesia alba ("albus" means "white") to distinguish it from "magnes," which is black. In 1831, the French chemist Antoine A. B. Bussy isolated a metal from a chemical related to magnesia alba and named it *magnesium*. And so ancient Magnesia ended by having two metals and an important natural force named after it.

Mammal

ALL MULTICELLULAR organisms are divided into two "kingdoms," the plant and animal. It was only in recent times that plants were recognized as being as fully alive as animals. Animals could breathe and move, which plants could not. In fact, the word *animal* comes from the Latin "anima" (breath).

In common speech, the word *animal* is often restricted to those with four legs and hair, such as dogs, cats, and cows, so that people will speak of "animals and birds." Actually, of course, a bird is an animal and so is an oyster, a butterfly, or an earthworm. The four-legged animals with hair (unlike those without hair, such as birds, reptiles, and insects) bear their young alive and feed them on milk formed in the mother's body. The milk-forming glands are called, in Latin, "mammae" and hence the hairy animals are called *mammals*.

There is one small group of primitive mammals, restricted to Australia and New Guinea, that, unlike other mammals, actually lay eggs. The best known of these has broad webbed feet like those of a duck and a snout that looks uncommonly like a duck's bill. It is therefore called the *platypus* (from the Greek "platys," meaning "broad," and "pous," meaning "foot") or the *duckbill* (derivation obvious) or the *duckbilled platypus*. It is also called the *ornithorhynchus* from the Greek "ornis" (genitive, "ornithos") (bird) and "rhynchos" (beak), or "bird-beak" which is just another way of saying duckbill.

The other egg-laying mammal (which occurs in several varieties) is the *spiny anteater*. (It has spines and eats ants so that is a good name.) It is also called the *echidna*, which makes less sense. In Greek mythology, Echidna was a monster that was half woman, half serpent. Perhaps the spiny anteater struck its discoverers as another kind of monster, half mammal (hair) and half reptile (eggs).

Because these creatures have one common exit for wastes (and eggs) they are called *monotremes* from the Greek "monos" (solitary) and "trema" (opening).

Marsupial

AUSTRALIA WAS cut off from the remaining land areas of the world at a time when the only living mammals were extremely primitive and were either egg-laying (see MAMMAL) or, if bringing forth young alive, were doing so at a very early stage of development. What's more, Australia remained cut off, so that these primitive mammals could develop and multiply, while in the rest of the world, more efficient mammals evolved and took over.

The best-known native mammal of Australia is the *kangaroo*. The story is that when the ship of the English explorer James Cook first landed in Australia in 1770 and the men saw a strange leaping animal, they asked the natives, "What's that?" The natives answered, "Kangaroo," meaning "What are you saying?"—and that was it.

In the kangaroos and related animals, the young are born after only a very short development. They emerge into the world with the bare ability to crawl through the mother's fur and into a special pouch where they feed on milk and remain until grown large enough to be able to assume independent life.

This pouch, within which the tiny newborn animals develop, is the most typical feature of the kangaroo and its relatives. The Latin word for "pouch" is "marsupium," so these animals are called *marsupials*.

The "higher" mammals allow their young to develop a longer time within the body before birth. In the case of man, the period of *gestation* (that is, the time during which the young are carried within the body before birth, from the Latin "gestare" meaning "to carry") is nine months; in elephants and whales, two years.

This is made possible by the fact that the mother bearing the child develops a special organ called the placenta, which is penetrated by blood vessels of both the mother and the developing baby so that food and air can cross in one direction and wastes in another. (The blood vessels approach one another but do not actually join.) The word *placenta* is a Latin one, meaning a kind of flat cake, which is what the organ looks like. The higher mammals are called *placental mammals*, for this reason.

Mathematics

MATHEMATICS IS that branch of knowledge that deals with quantities, their measurement, and their interrelationship. It is derived from the Greek "mathein" (to learn), which implies that it is the subject which, above all others, must be learned. We admit that, partially, ourselves, when we speak of the "three R's" as the absolute fundamentals of learning: "readin' and 'ritin' and 'rithmetic, taught to the tune of the hickory stick."

And *arithmetic* (from the Greek "arithmos," meaning "number") may well have come first. People, even in advanced cultures, can get along (though rather poorly) if they cannot read or write, but even illiterates find it useful to add. Some primitive tribes are supposed to have no names for numbers over two, but any tribe that gets to the stage of making stone axes must want to count higher than that, if only to make sure no one has been appropriating an ax that wasn't his.

Mathematics is generally not counted among the natural sciences (see SCIENCE). It is admitted to be the most important tool of many of the sciences, but the point is made that a "natural science" begins with the observation of nature and the collection of facts. These are then brought into order by the use of mathematics. Mathematics, itself, however, is apart from nature and is a creation of pure mentality, being built up step by step by the mathematician with closed eye and thoughtful mind from a minimum number of fundamental and obvious principles called *axioms*. This comes from the Greek "axios," meaning "worthy," perhaps because axioms are principles which all people will agree are "worthy" to serve as foundations.

An example of an axiom is: A straight line is the shortest distance between two points.

One can't help but wonder how axioms arose, in the first place. Did they spring up, pure and unsullied, in the mind? Or did some early thinker observe two points and draw various connecting routes before deciding by observation and experiment that it looked as though a straight line were probably the shortest connecting route between them?

Melanin

IN MAN the most noticeable natural colors are those of the blood and those of the hair, skin, and eyes. The colored substance of blood, as it happens, is not named after its color (see PORPHYRIN; HEMOGLOBIN), but that of the hair, skin, and eyes is another story. (Those three are lumped together because the same substance is involved in each.)

The color of hair, skin, and eyes is due mostly to a blackish pigment, produced in varying amounts by practically all human beings. This pigment is called *melanin* from the Greek "melas" (black). The few people who cannot form this substance have very fair skins, white hair, and eyes that would be colorless except that the blood vessels show through, giving them a pink appearance. Such people are *albinos*, from the Latin "albus" (white).

People who can form a little melanin are yellow-haired and blue-eyed. (The blue of blue eyes is not the result of a blue pigment but the effect of tiny particles of melanin which scatter blue light the way dust particles in the air scatter it to form the blue sky.) Such people are *blond*, a word which is of ancient German origin meaning "light" or "fair."

People who can form a fair amount of melanin have brown hair (or even black hair) and brown eyes. There is enough melanin in the skin to give them a swarthy complexion. (*Swarthy* comes from the Anglo-Saxon "sweart," which is analogous to the German *schwarz*, meaning "black.") Such people are *brunet*, a French word originating from "brun" (brown).

If the amount of melanin formed is high enough, the skin itself is distinctly brown, sometimes quite dark brown. The African Negro and his descendants are an example of this. The word *Negro* is a Spanish word meaning "black," derived from the Latin "niger" (black). (The Spaniards and Portuguese were the first of the modern Europeans to make contact with the dark-skinned inhabitants of West Africa, which is why a Spanish word has come into use in this connection.)

Mercury

THE ANCIENTS knew seven metals: gold, silver, copper, iron, tin, lead, and mercury, and of these mercury was by all odds the most remarkable. It was the only liquid among them, and it was such a heavy liquid, too; only gold was heavier. The Greeks called the metal "hydrargyros" from "hydor" (water) and "argyros" (silver); i.e., "liquid silver." That name still lingers today in the chemical symbol for mercury, which is Hg.

The notion of mercury's being a special kind of silver lingers also. Unlike ordinary silver which just lies there in a lump, mercury, being liquid, moves and vibrates and, if spilled, breaks into thousands of silvery droplets that race about. There is something seemingly alive about it and one of its names in English is *quicksilver*, where *quick* has its older meaning of "alive" as in the phrase "the quick and the dead."

The medieval alchemists, who were great for masking what knowledge they had in mystical gobbledygook, were struck by the fact that there were seven known metals and seven known planets. They thought there might be some mysterious significance to this and matched up the two sets.

Gold they referred to as Sol (Latin for "sun") and silver, Luna, (Latin for "moon") because they were the most precious and should be named after the most prominent "planets." Copper, as next most precious, was named Venus which was next most prominent. Iron was a natural for Mars because of the use of iron weapons in war, while the heavy and dull lead seemed appropriate for the slow-moving Saturn. Tin was Jupiter, by elimination, probably, since hydrargyros, the one remaining metal, could only be tied up with one planet, Mercury.

Just as hydrargyros was the most alive of the metals, being liquid and mobile, so Mercury was the most alive of the planets since its apparent motion across the sky was most rapid. So the liquid metal received the name *mercury*, for, after all, Mercury was the wing-shoed messenger of the gods, fleet as thought, and of all the planet-named metals, only that one kept its planet name — until the fashion was begun again in modern times (see URANIUM).

Meridian

Any line drawn due north and south on the surface of the earth (or against the vault of the sky), passing from pole to pole, is called a meridian. Wherever you are, you are standing on (and under) one such line. About noon (time by clock can vary several minutes from the time marked by the sun), the sun passes across the meridian on which or under which you are standing, and that is the middle of the day for you, exactly halfway between sunrise and sunset. The Latin word "medius" is "middle" and "dies" is "day"; consequently "medidies" means "midday." "Medidies" was corrupted into "meridies" and from that comes the word *meridian*.

(Strangely enough, the word *noon* comes from the Latin "nonus" (nine). Originally, it referred to the ninth hour of the day counting from sunrise which, on the year-average, is at 6 A.M. This put "nonus" at 3 P.M. or halfway between midday and sunset. Thus, what was originally the middle of the afternoon has come to mean the middle of the day itself.)

The position of a particular point on earth's surface can be given, at least in an east-west sense, by choosing some particular meridian as a zero point, then marking off the rest of the earth in equally spaced meridians.

At first, most nations chose their own capital (or some other point particularly convenient to themselves) as starting point and figured east and west from that. This aroused confusion in sea travel, where a knowledge of accurate position was more important than on land, and where ships of all nations, each with its own system, found things a mess.

In the middle nineteenth century, Great Britain had more ships and a larger seagoing trade than any other. It was natural, then, for the Washington Meridian Conference in 1884 to adopt Great Britain's standard internationally. The meridian passing through the Greenwich Astronomical Observatory in London is accepted as the zero point the world over now, and that is the *Prime Meridian* (from the Latin "primus," meaning "first").

Metal

For hundreds of thousands of years, mankind had been using stone and wood for various tools. Wood was easier to handle but stone could take more punishment. Both were useful.

And then, not more than six thousand years ago, men discovered a new kind of material altogether. Perhaps they first found small nuggets of yellow gold. Or perhaps somebody built a charcoal fire on some green rock in the Sinai peninsula and found ruddy drops of copper under the ashes afterward. In any case, copper, gold, and silver date back to nearly 4000 B.C.

The qualities that struck men first about these substances must have been those in which they differed from stone and wood. Take gold as an example. It reflects light beautifully and shines, instead of having the dull appearance of stone. It shows *luster*, from the Latin "lucere" (to shine).

Gold can also be hammered into various shapes or into very thin sheets, whereas stone so treated will simply powder. Gold is therefore *malleable*, from the Latin "malleus" (hammer).

Gold can be drawn out into a thin wire, whereas again stone, so treated, would break and powder. Since one end of a heated strip of gold will seem to follow the pincers that are drawing it out and leading it, so to speak, away from the other end, gold is said to be *ductile*, from the Latin "ducere" (to lead).

Then, too, gold sheets or gold wire can be bent into any shape without breaking. (Can you imagine stone bending?) Gold is therefore said to be *flexible*, from the Latin "flectere" (to bend).

For all these reasons, appearance plus the way these substances could be bent and drawn and hammered, they were ideal for the fashioning of jewelry and ornamentation, so they were eagerly desired even before more practical uses turned up. Since they were much rarer than wood or stone, they had to be carefully searched for. The Greek word "metallon" means "mine" and also "metal." But this probably comes from the Greek word "metallan," meaning "to search for." Anyway, the new class of materials received the name of *metal*.

Meter

In 1791, the French nation, in the midst of a revolution, wished to break with the past, especially with those aspects of the past which they considered illogical and useless. For instance, the system of weights and measures then in traditional use was overly complicated and, moreover, varied from place to place.

They began by trying to set up a unit of distance equal to one forty-millionth part of the earth's circumference. Unfortunately later measurements showed that the unit they had chosen was not exactly that fraction and today the unit is defined simply as the distance between two marks on a platinum-iridium bar kept in Sèvres, a suburb of Paris. This unit they called the *mètre* (*meter* in English), from the Latin word "metrum," meaning "measure." The whole system of measurements which began with the meter is called the *metric system* and is today used by scientists the world over.

The metric system is built up in units of ten, using Greek prefixes for multiples and Latin prefixes for fractions. Thus, ten meters is a *dekameter*, a hundred meters is a *hectometer*, a thousand meters is a *kilometer*, and ten thousand meters is a *myriameter*—from the Greek words "deka" (ten); "hekaton" (hundred); "chilioi" (thousand), and "myrioi" (ten thousand).

On the fraction side, a tenth of a meter is a *decimeter;* a hundredth of a meter is a *centimeter*, and a thousandth of a meter is a *millimeter* —from the Latin words "decem" (ten); "centum" (hundred); and "mille" (thousand).

The Greeks had no word for a number larger than ten thousand and the Romans no word for a number larger than one thousand. However, the system was extended by using less specific words. For instance, a *megameter* is one million meters, from the Greek word "megas," meaning simply "large"; while a *micrometer* is a millionth of a meter, from the Greek word "mikros," meaning simply "small."

In 1958, new prefixes were adopted internationally. The prefix "giga-" from the Greek "gigas" (giant) stands for a billion and "tera" from the Greek "teras" (monster) stands for a trillion, so that a *gigameter* is a billion meters and a *terameter* is a trillion meters. On the other hand, "nano" from the Greek "nanos" (dwarf) stands for one billionth and "pico" for one trillionth, so that a *nanometer* is a billionth of a meter and a *picometer* is a trillionth of a meter.

Microbe

One Dutchman first used lenses to penetrate the incredibly distant (see telescope); another Dutchman, Anton van Leeuwenhoek, used lenses to penetrate the incredibly small. Leeuwenhoek produced lenses so powerful in magnification that he could see single cells. These were the first *microscopes*, from the Greek "mikros" (small) and "skopein" (watch). With them, Leeuwenhoek indeed "watched the small."

In 1675, he described *protozoa*, from the Greek "protos" (first) and "zoon" (animal), which are single-celled animals and were indeed the "first animals" to exist on earth. In 1683, he discovered single-celled organisms smaller still, which were not animals nor, yet, quite plants.

These are today known by a variety of names. The most general name is *germ* (from the Latin "germen," meaning "bud," i.e., something small that yet contains the beginning of life in it). *Germ* is also the name of cells from which complex organisms develop. The human ovum and sperm are, for instance, referred to as *germ cells.* The part of a plant seed that develops into the plant (i.e., that *germinates*) is also a germ, so that we may speak of *wheat germ.*

Another name for the small creatures is *microbe.* The "-be" ending is what is left of the Greek "bios" (life). But "small life" really includes the protozoa, too, and still other forms. Because *microbe* has been applied so often to just a particular type of cell, the modern term for "small life" in general is *microorganism.*

The name most used today for Leeuwenhoek's creatures is *bacteria,* from the Greek "bakterion" (a little rod), since a number of them do indeed have the appearance of tiny rods. The study of such creatures is now known as *bacteriology*. (The suffix "-logy" comes from the Greek "logos," meaning "word"; a bacteriologist is someone who speaks words about bacteria.) The study of microorganisms in general is *microbiology* ("words about small life").

1,000,000,000,000... Million

THE LARGEST number-word among the Romans was "mille" meaning "thousand." It has come down to us in a number of ways. A millipede is an insect with a thousand feet (see CENTIPEDE) and a millimeter is a thousandth of a meter (see METER). The Romans used as a measure of length the distance covered by one thousand paces of their marching legions. This they called a "milia" and it has come down to us as *mile.*

In medieval Italy, with the growth of finance and the increasing prosperity of Europe, it became handy to have words for numbers larger than a thousand. So for the number 1,000,000 (a thousand thousand), someone tacked the ending "ione" on to "mille." This suffix implied something of unusual size so that "millione," or *million* in English, is a "king-size thousand," so to speak.

In the fifteenth century, still larger words were needed and the French invented *billion* for a thousand million (1,000,000,000). The ending "-llion" had already become established as implying a large number and the "bi-" prefix came from the Latin "bis" (twice), perhaps because *billion* was the second made-up word of the sort.

By using additional Latin number prefixes, still larger numbers could be invented. *Trillion* is a thousand billion ("tres" is Latin for "three") and *quadrillion* is a thousand trillion ("quattuor" is Latin for "four"). This can continue indefinitely.

The French use these names for numbers at multiples of a thousand, England and Germany at multiples of a million. Thus a quadrillion is a thousand trillion in France, but a million trillion in England or Germany. America went along with the French system after the Revolutionary War when France was popular with us and England was not. Oddly enough, the French dropped the word *billion.* They now call 1,000,000,000 a *milliard.* The suffix "-ard" again implies something has gone all the way (a drunkard is one who drinks all the way) so that 1,000,000,000 is a "very king-size thousand," so to speak.

Molecule

THE LATIN word "moles" means "a mass" and the diminutive form "moleculus" naturally means "a small mass." The term *molecule*, was therefore originally used to mean any tiny particle of matter.

In the study of gases, however, it early became obvious that they must be composed of tiny flying bits of matter — molecules — separated by gaps. In 1811, moreover, the Italian physicist Amadeo Avogadro suggested that a given volume of any gas at the same temperature and pressure contained the same number of molecules. This was *Avogadro's Law*, which showed that one could determine the comparative weights of different molecules by measuring the densities of gases.

It turns out that the molecules of some gases (e.g., helium and argon) consist of single atoms, but that in most cases, gas molecules consist of two or more atoms. The molecule concept has been extended to liquids and solids also, and a molecule is now defined as the smallest particle into which a compound can be subdivided without change in chemical properties. Molecules are now known that are made up of millions of atoms.

The *molecular weight* of a substance is the sum of the weight of the individual atoms composing its molecules. Each atom has an *atomic weight* relative to that of the oxygen atom which is set, arbitrarily, at exactly 16. According to this scale, carbon weighs just a bit over 12 and hydrogen a bit over 1. A molecule of ethyl alcohol (with two carbon atoms, six hydrogen atoms and an oxygen atom) has, therefore, a molecular weight of 24 plus 6 plus 16, or 46.

If the molecular weight is expressed in grams, then the result is a *gram-molecular weight*, which is usually abbreviated as *mole* (or sometimes *mol*). Thus, 46 grams of ethyl alcohol (46 being its molecular weight) is the equivalent of one mole of ethyl alcohol. It is convenient for chemists to talk about moles rather than grams sometimes, because it turns out that one mole of any substance contains as many molecules as one mole of any other. The number of molecules in a mole is 602,000,000,000,000,000,000,000 (or, *Avogadro's number*, since the existence of this number was deduced from Avogadro's Law).

Moment

A SEESAW is a flat board supported in the center. The support is the *fulcrum* of the seesaw, from the Latin "fulcrum" (a bedpost), which in turn comes from "*fulcere*" (to prop up).

If two children of equal weight sit at opposite ends of the board, and one pushes himself up, the other will come down; the other pushes and the situation reverses itself. The downward pull of gravity on one child is transmitted as an upward lifting force on the other. The seesaw is thus an example of a simple *machine*, which may be defined as any device that will allow force applied at one point to have its effect at another point or in another direction. Machines are generally used as a means of doing work that would be more difficult without such transmission or direction-change of force; the word is derived from the Greek "mechos" which means "means" or "expedient."

But suppose one child on the seesaw is twice the weight of the other. The lighter child, once in the air, will not be heavy enough to lift the other. The seesaw will come to a halt. In order to rebalance the seesaw, it is necessary for the heavier child to sit closer to the fulcrum. The ability of a force to turn a board about a fulcrum depends not only upon the size of the force but also upon the distance from the fulcrum of the point at which it is applied. For a well-balanced seesaw, the weight of one child times his distance from the fulcrum must equal the weight of the other times his distance.

When force times distance on one side of the fulcrum is equal to force times distance on the other, we say the *torques* are the same. (The word comes from the Latin "torquere" meaning "to twist.") Or we can say the *moments* are the same. This comes from the Latin word "momentum" which, in turn, is just a shortened version of "movimentum" meaning "movement." The two words refer to the "twisting" or "moving" effect of a force.

The word *moment* has also come to mean an instant of time. Since time is always measured by movement (of heavenly bodies, of pendulums, etc.), a small "movement" in space is a small "moment" of time.

Monosaccharide

Proteins, the important giant molecules of living tissue, are built up out of long strings of relatively small molecules called amino acids. In the process of digestion, these long strings are broken up into smaller strings. Since the Greek word for "cooking" or "digestion" is "pepsis," the short strings of amino acids resulting from digestion were named *peptides.*

Different peptides may be distinguished according to the number of amino acids in the chain. A peptide made up of two (or three or four) amino acids is called a *dipeptide* (or *tripeptide* or *tetrapeptide*) using Greek number prefixes (see neoprene). Proteins themselves, since they contain a large and indefinite number of amino acids (sometimes thousands) are *polypeptides,* from the Greek "polys" (many).

There are a number of kinds of giant molecules in living tissue but in every case they are built up of smaller units joined together. For instance, starch is made up of a number of glucose units (see glucose) joined in a string. Starch is therefore an example of a *polysaccharide* (many-sugar), since glucose is a kind of *sugar* and the Latin "saccharon" (from the Greek "sakchar") means "sugar."

Glucose itself is a *monosaccharide* ("one-sugar") the Greek "monos" meaning "one." (However, there is no analogous "monopeptide" since a single amino acid is just called an amino acid. Even scientists are not always logical.) Two other important monosaccharides are *fructose* and *galactose;* each has the same number of the same kind of atoms as glucose, but the atoms are arranged differently. Fructose is found in a variety of sugar contained in fruits (the Latin "frux" means "fruit") while galactose is found in the variety of sugar contained in milk (the Greek "gala" means "milk").

Glucose and fructose will combine to form a *disaccharide* ("two-sugar") called *sucrose.* This is directly from "saccharon" since it is common everyday sugar, the kind we buy in the store. Glucose and galactose form a disaccharide called *lactose,* the sugar in milk (the Latin "lac" means "milk").

Mutation

DURING THE process of cell division, each chromosome (see GENE) forms a duplicate of itself. When cell division is complete and there are two cells where only one existed before, it is found that the original of each chromosome is in one cell and the duplicate in the other. Since the duplicate is usually identical with the original and since the chromosomes govern the chemistry of the cell, the new cell is just like the old one.

The same principle governs the formation of the sex cells and the growth of the new individuals arising from them, so that giraffes have baby giraffes and elephants have baby elephants rather than vice versa.

However, the duplication is not always perfect. Whenever, through some chance, a new chromosome is formed that is not exactly like the old, the new cell is not exactly like the old, and the new young creature is perhaps not exactly like its parents. In 1886, the Dutch botanist Hugo de Vries noted a group of plants in which some were radically different from the rest, though all seemed to have originated from the same stock. He applied the word *mutation* to such a sudden change in going from parent to offspring, from the Latin "mutare" (to change).

Actually, mutations are an old story to herdsmen. Domestic animals (and human beings, too) sometimes give birth to young that are so far out of the ordinary as to seem *freaks* of nature. The word comes from the Anglo-Saxon "frician" (to dance). A wild dance is characterized by startling changes in step and movement, so the word comes to mean any unexpected chance happening (as a freak wind or a freak bounce of a ball).

Freaks appear to be a kind of practical joke perpetrated by nature. It is almost as though nature is making grim sport of the herdsman expecting a normal calf or lamb, or even grimmer sport of a human mother, and freaks are also, in point of fact, called *sports.* On the other hand, a freak birth was considered an ominous warning by the diviners of ancient Rome, and from the Latin "monere" (to warn), they got "monstrum" (a divine warning of ill omen) and we get our word *monster.*

Neoprene

THE FIVE-CARBON compound isoprene (see TERPENE) is used by living tissue (particularly plant tissue) as a unit out of which to build up a number of more complicated compounds. A molecule made up of two isoprene units is a terpene. Similarly, one made up of four isoprene units is a *diterpene* ("two-terpene," from the Greek "dyo," meaning "two"); one of six isoprene units is a *triterpene* ("three-terpene," from the Greek "treis," meaning "three") and one of eight isoprene units is a *tetraterpene* ("four-terpene," from the Greek "tettares," meaning "four"). A molecule made up of three isoprene units is a *sesquiterpene*. The Latin prefix "sesqui" is a running together of "semis" (half) and "que" (and) and means one "and a half."

Certain important compounds are built up out of isoprene. Vitamin A, for instance, is a diterpene and carotene (see CHLOROPHYLL) is a tetraterpene. However, the most familiar substance built up out of isoprene units is rubber. The rubber molecule is a large one, made up of an indefinite, but high, number of such units. It is a *polyterpene* (or "many terpenes," from the Greek "polys" meaning "many").

As rubber grew more valuable, chemists tried to prepare artificial rubber in the laboratory. They began by working with isoprene, trying to force it to build up large molecules like those in rubber. Until recently, they failed. The isoprene molecules joined all right, but not in the correct fashion.

They had better luck with compounds similar to isoprene. One was a compound with a molecule in which a chlorine atom replaced the one carbon atom branching off the main carbon chain in isoprene. This compound was named *chloroprene* (a combination of *chlor*ine and is*oprene*). In 1931, the Du Pont Corporation successfully manufactured an artificial rubber built up out of chloroprene units. At first, they called it Duprene, in honor of the company, but then they changed the name to *Neoprene*, the Greek prefix "neo-" coming from "neos" (new), and so the nonsense syllable "prene" (see TERPENE) has really branched out.

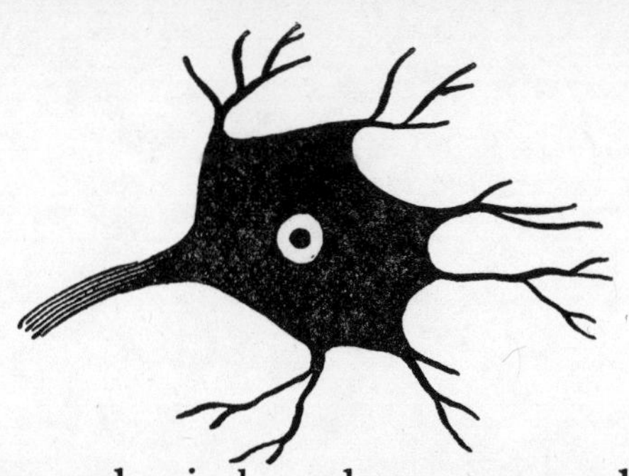

Neuron

The brain and spinal cord are composed of irregular cells that usually have a number of short branching fibrous extensions at one end and a single long fibrous extension at the other. The long fibrous extension is encased in a fatty sheath and this is by far the most noticeable part of the cell. Without a microscope, in fact, it is the only part that can be noticed.

The Romans used the word "nervus" to apply to any fiber occurring in the animal body — a tendon, sinew, or this extension of a brain cell. With time, the word was used only for the last, which is now called a *nerve*.

By 1891, the German anatomist Heinrich Waldeyer, who first studied the nerve cells thoroughly, wanted to get away from *nerve* and its implication of the long fiber only, so he turned to the Greek equivalent, *neuron*, for the entire cell. (Of course, the Greek word also meant tendons and sinews originally, but in modern anatomy, Waldeyer's meaning stuck.) It is from "neuron" that we get words like *neuralgia* for pain along the course of a nerve (the Greek "algos" means "pain"); *neuritis* for inflammation of a nerve (the Greek suffix "-itis" means "inflammation"); *neurology* for the study of the nervous system and its diseases (the Greek "logos" means "word" and studies always result in many "words" in lectures and books); and *neurosis* for a mental ailment or nervous condition (the Greek suffix "-osis" means "a condition of").

The small branched extensions at one end of the neuron are the *dendrites*, from the Greek "dendron" (tree) which is what the branches resemble. The long extension, running as it does down the center of the fatty sheath, was called the "axis" at first (see VERTEBRA), but this was later modified to *axon* to have it match the "-on" suffix of *neuron*.

The axon of one nerve cell branches at the end and these branches usually mingle with the dendrites of another, though they do not join. Nerve impulses can cross this microscopic gap, which is called a *synapse*, from the Greek "syn-" (together) and "haptein" (to fasten); it is at this point that nerves are "fastened together."

Niacin

THE MOLECULE of nicotine (see NICOTINE) consists of two rings of atoms. It was found, in 1867, that one of the rings could be destroyed by treatment with strong acid. What was left showed acid properties and was therefore named *nicotinic acid*. The resemblance in names does not extend to properties. Nicotine is a violent poison, but nicotinic acid is not.

Now to change the subject. A certain disease marked by mental difficulties, a sore mouth, and an inflamed roughened skin has long been prevalent in Spain, Italy, and the southern parts of the United States. The disease is named *pellagra*, from the Italian "pelle agra" (rough skin). In 1915, J. Goldberger, a physician working for the United States Public Health showed, by experiments on prisoners, that he could cause the disease in healthy people by limiting the diet, and then cure it by adding such foods as milk to the diet. Pellagra, like scurvy (see ASCORBIC ACID), was a vitamin deficiency disease and not a germ disease. Temporarily, the vitamin that prevented pellagra was called the *P-P factor* (the P-P standing for "pellagra-preventive").

Then, in 1937, the American biochemist C. A. Elvehjem showed that *nicotinamide* (with a molecule consisting of nicotinic acid to which an amine group was attached) was the compound that would prevent pellagra. A few months later, he showed that nicotinic acid itself would do the job, too, since the body could convert it to nicotinamide easily.

This presented doctors with a problem. The name nicotinic acid could be easily confused by the general public with the name nicotine. People might get the idea they could get vitamins by smoking. Or a person, seeing a label on a food product stating that it contained nicotinic acid, might think it contained poison.

It was therefore decided to take the first two letters of *nicotinic* and the first two of *acid*, add the suffix "-in" by analogy to "vitamin," and end up with the made-up word *niacin* to prevent such confusion. And nicotinamide became, in the same fashion, *niacinamide*.

Nicotine

The tobacco plant is a native of the western hemisphere, and Europe did not see it until samples were brought into Spain from America in 1558. Two years later, the French ambassador to Portugal, Jean Nicot, sent tobacco seeds to Catherine de Medici, mother of the French king. In Nicot's honor, the group of plants to which tobacco belonged was eventually given the Latinized name of *Nicotiana.*

Tobacco is one of a number of plants that get rid of some excess nitrogen by forming complicated nitrogen-containing organic compounds and storing them. These compounds possess weakly alkaline properties (see potassium) and are therefore called *alkaloids.* The suffix "-oid" means "having the form of" from the Greek "oeides" (form).

Often the alkaloid is named after the Latin name of the *genus* (a Latin word meaning "group" or "kind") to which the plant belongs. Invariably, their names are given the "-ine" suffix which chemists reserve for nitrogen-containing organic compounds. The chief alkaloid of tobacco received, therefore, the name *nicotine.*

Other familiar alkaloids named in this fashion include *strychnine,* which is found in *nux vomica,* the seed of an Indian tree which belongs to the genus Strychnos. *Coniine* is found in poison hemlock and is, therefore, the poison that killed Socrates. Hemlock belongs to the genus Conium.

Sometimes alkaloids are named for other reasons. *Quinine*, for instance, the world's first effective medicine for malaria, is found in the bark of a tree that originally grew in South America. The South American Indian word for "bark" was "quina," hence the name. *Morphine*, which is obtained from the opium poppy, can be useful, under a physician's guidance, in relieving pain and lulling a suffering patient to sleep. Its name comes from Morpheus, the Roman god of sleep. (However, the opium poppy belongs to the genus Papaver — the Latin word for "poppy" — and another alkaloid from that plant, which contains a number of them, is named *papaverine* in consequence.)

Niobium

In recent years, the United States has received the honor of having an element named after it (see GALLIUM). It had, however, received the honor more than a century ago and lost it.

It came about in this way. John Winthrop the Younger, Governor of Connecticut in 1635, was an amateur mineralogist and came across a fragment of strange rock near his home in New London. Eventually, this rock was sent by his grandson to London (where it is still preserved in the British Museum). In 1801, the English chemist Charles Hatchett detected in the rock element number 41, which he named *columbium*, in honor of the country in whose territory the mineral was first discovered. (The United States, or "Columbia," as it is sometimes called in patriotic poems, was by then an independent nation.)

However, this did not end the story. In 1802, the very next year, the Swedish chemist Ekeberg (see TANTALUM) discovered the element tantalum. Columbium and tantalum are very similar chemically and, in 1809, the English chemist William Hyde Wollaston decided that the two were identical, and the chemical profession, in general, went along with the notion.

Of course, if they were the same element, Hatchett was still first by a year and his name should have stuck, but Berzelius, Europe's leading chemist, thought Ekeberg's work more thorough and convincing, and in 1814 voted against columbium and for tantalum and again the rest of the profession followed.

Finally, in 1846, the German chemist Heinrich Rose showed columbium and tantalum to be two different elements. However, because of their similarity, Rose called columbium *niobium*, after Niobe, the daughter of Tantalus (see TANTALUM).

For many years, the element kept its two names, columbium in America and *niobium* in Europe. Recently, however, an international conference of chemists has decided to make niobium the official name of the element everywhere, and America lost the honor of the name.

Nitrogen

It was in the 1770's that chemists first realized there were two substances in air, one of which was necessary to life, and one of which could not support it. If animals were forced to remain in a closed container of air, or if wood were burned in one, the part of the air that was essential to life was used up. Substances would not burn in what was left, and animals would not live in it.

As a result, this gas received a number of unpleasant names. The Swedish chemist Karl Wilhelm Scheele called the life-supporting fraction fire air, and the remainder foul air. (Scheele's "fire air" was oxygen, which he discovered two years before Priestley (see OXYGEN). Priestley gets the credit, though, because his results were published at once, while Scheele's were delayed.)

The British chemist Daniel Rutherford in the same year, 1772, named the "foul air" fraction *mephitic air*, from the Latin word "mephitis," meaning "poison gas." The French chemist Antoine-Laurent Lavoisier called it "azote," from the Greek prefix "a-," meaning "not," and "zoe," meaning "life." Azote was therefore the gas that was "without life." The Germans call it, on the same principle, Stickstoff, which means, in German, "suffocating substance."

None of these names stuck in English, although "azote" gave rise to a number of names in organic chemistry for various substances that contain nitrogen atoms. For instance, there are azo compounds, diazo compounds, hydrazo compounds, azoxy compounds, and so on.

In 1790, the French chemist Jean-Antoine Chaptal did the trick. The new substance was found to form part of the molecule of the common chemical known as niter (one of the components of gunpowder). Since there was a fashion in those years to name new gases with the ending "-gen," from the Greek suffix "-genes," meaning "born" or "produced," Chaptal called the gas *nitrogen*. It was something, in other words, from which niter was produced.

Notochord

In the human, the skeleton develops in stages. In the early embryo, it begins as a straight rod of cartilage down the middle of the back. In 1848, the English physician Richard Owen named this the *notochord* from the Greek "notos" (back) and "chorde" (string). The rest of the skeleton — first of cartilage, then of bone — develops in stages, the process of bone formation not being complete until some years after birth. This happens not only in man, but in all mammals, birds, reptiles, amphibians, and bony fishes. In sharks and their relatives, the skeleton remains cartilage throughout life (see TELEOSTEI).

In some very primitive animals, the skeleton never passes the notochord stage. There is, for instance, a little animal about two inches long, vaguely fishlike, whose only skeleton is a notochord running the length of its back. Its front and rear ends both end in similar pointed fashion, so that at a quick glance it is hard to tell which is front, which rear. It is therefore named *amphioxus* from the Greek "amphi" (both) and "oxys" (sharp); it is "sharp at both ends."

There are animals still more primitive which never even develop a full notochord, only the beginnings of one; or develop one in early life and lose it in adulthood. There are the *tunicates*, for instance, so called because in adult form they are covered with a hard layer or "tunic." They live as motionlessly as oysters and have no trace of any internal skeleton. In fact, for many years they were classified as mollusks, the group to which clams and oysters belong. The young tunicate, however, fresh from the egg, swims about freely, and has a notochord.

To a zoologist, the development of a notochord, however imperfect or temporary, shows a relationship to other animals with internal skeletons — a relationship that animals such as lobsters and beetles, which never form the slightest trace of internal skeleton at any time, do not have. Consequently, all animals that make even a beginning at a notochord, including the tunicates and man, are put in the phylum *Chordata*.

Nova

In 1572, a very unusual thing happened in the heavens. Among the stars, there appeared a new one, in the constellation Cassiopeia. It grew so bright it could be seen in daylight, and then it faded away. To a world which was used to considering the heavens perfect and unchanging (see ALGOL and COMET) this was a noteworthy thing. One observer was a young Danish scientist named Tycho Brahe who, the next year, published his observations under the title *De Nova Stella* ("Concerning the New Star"). And from that day a "new star" is a *nova* (the feminine form of the Latin "novus," meaning "new").

In 1604, another nova appeared and was this time observed by Brahe's pupil Johann Kepler, who also wrote a book with a title *De stella nova in pede Serpentarii.*

Of course, novae, as we now know, are not new stars but old ones that have exploded. There are about 25 novae occurring every year in our own galaxy. Most of them are unspectacular, stars that increase a few hundredfold in brightness, then fade. (Of course, if the sun were to do this, life on earth would be destroyed, but astronomically such a phenomenon is still rather mild.)

Occasionally, though, a star tears itself apart completely and increases in brightness a few hundred-billionfold and outshines an entire galaxy of stars. Such a star is a *supernova* (the Latin "super" meaning "above" or "beyond").

Only three supernovae are known to have occurred in our own galaxy. One was Brahe's nova and one Kepler's. Two supernovae, in other words, occurred in a single generation, and none since, to the great frustration of astronomers who now have telescopes and cameras.

The third supernova occurred in 1054, but it was observed by Chinese and Japanese astronomers only. At the point indicated by them, there now exists a small mist of light that is obviously the cloudy remnant of a super-explosion which because of its shape is called the *Crab Nebula* (see GALAXY for meaning of *nebula*).

Nuclear Reactor

THE FIRST self-sustaining nuclear reaction (see NUCLEUS) was set up at 3:45 P.M. of December 2, 1942, under the stands of a football stadium at the University of Chicago. The reaction took place within a huge cube of uranium and carbon. The uranium atoms were split by slow-moving neutrons and that (see FISSION) liberated the energy. The carbon was needed to slow the neutrons to the proper speed and was therefore the *moderator*, from the Latin "moderare" (to regulate) since it "regulated" the neutron speeds.

Because the cube was made by first forming a layer of uranium, then piling a layer of carbon on it, then uranium on that, then carbon, and so on, it was called an *atomic pile*. For a while, all devices in which a self-sustaining nuclear reaction could take place were called piles. However, since later devices were formed in less makeshift fashion, they began to be called, much more properly, *nuclear reactors*.

The adjective *nuclear* is better than *atomic* anyway, if we are talking about nuclear reactions. Thus, a bomb fired through the energy of uranium fission was called an *atomic bomb*, in the first announcement by President Truman that such a device had been dropped on Hiroshima in August, 1945. The newspapers shortened this to *atom bomb* and to *A-bomb*, but all are really misnomers. An ordinary TNT bomb involves atomic reactions and could be called an atomic bomb. The thing about the A-bomb is that it involves nuclear reactions; it should, therefore, be called a nuclear bomb.

Again, a submarine run by a nuclear reactor is called an *atomic* submarine and said to be running on atomic power or atomic energy; but it should be *nuclear* submarine, nuclear power, and nuclear energy. The date December 2, 1942, mentioned above is even said to be the beginning of the *atomic age*, though we have been in the atomic age for thousands of years. It is the *nuclear age* we are now in.

But all this is useless. As in many other scientific names, the mistake has been made and it is probably already too late to correct it.

Nucleic Acid

In a series of experiments, beginning in 1869, a Swiss investigator named Friedrich Miescher discovered an acidic substance in the nuclei (see protoplasm) of cells. Because of its place of origin, he named it *nuclein.* The "-in" ending seemed to imply a protein, which it wasn't, so in 1889, it was renamed *nucleic acid.*

But apparently one change was all the chemical profession could stand for. When, some time later, "nucleic acid" was discovered in the cytoplasm of the cell, no effort was made to change it again. It remained "nucleic acid" whether in or out of the nucleus.

There was one difference, though. The nucleic acid of the cytoplasm contained a kind of sugar called *ribose* in its molecule. Before its discovery in this nucleic acid by the American biochemist P. A. Levene, in 1908, it had not been known to occur in nature. It had been synthesized by the German biochemist Emil Fischer in 1901, and he showed that it was very similar in molecular structure to another sugar called *arabinose.* Arabinose did occur in nature, in a kind of dried tree sap called *gum arabic*, whence its name. (Gum arabic was first imported from Arabia, and *gum* comes from the Latin "gummi" which comes from the Greek "kummi," meaning "sap.") In naming the new sugar, Fischer shuffled the letters in *arabinose*, left out a few, and came up with *ribose.*

The nucleic acid of the nucleus contained a sugar almost like ribose but with one oxygen atom missing. The usual way of naming such an oxygen-short compound was to add the prefix "deoxy-," the Latin prefix "de-" having as one of its meanings "something taken away." So the name was *deoxyribose.* Americans added an *s* in order to improve the sound and called it *desoxyribose*, but in 1956, an international convention settled the matter in favor of the *deoxy*-prefix.

Because of the nature of the sugar, the nucleic acid of the cytoplasm is *ribonucleic acid*, abbreviated commonly as *RNA*, while that of the nucleus is *deoxyribonucleic acid*, or *DNA.*

Nucleus

THE WORD *nucleus* (Latin for "a little nut" from "nux" (genitive, "nucis") meaning "nut") has been used to describe a small body in the center, more or less, of a plant or animal cell (see PROTOPLASM). It has been used for all sorts of objects at the center of a larger mass, or for some small mass out of which a larger mass develops. The word has found its place in atomic physics, too.

About 1906, the New Zealand-born British physicist Ernest Rutherford began to study the effect of firing alpha particles (see ALPHA RAYS) at thin sheets of metal. The vast majority of these particles went through as though nothing were there, but some were deflected and a very few even bounced back. Rutherford decided, therefore, that most of the mass of an atom was concentrated in the center, and this turned out to be the case.

The protons and neutrons (see PROTON), which are the heavy particles in an atom, are all concentrated in a very small central region of the atom, while all the outer regions are taken up by the very light electrons (see ELECTRON). Alpha particles plow through the electrons with no trouble but every once in a while one will strike the tiny central portion and, as Rutherford observed, bounce back. This central portion is called the nucleus, too. To distinguish it from the structure in the cell (the *cell nucleus*), it is called the *atomic nucleus*.

Ordinary chemical reactions — burning, rusting, all the activity that goes on in test tubes and living tissue — involve only some of the outermost electrons of the atoms involved. Moderate amounts of energy are given off or taken up in the process, from a match flame, to body heat, to a dynamite explosion.

Radioactivity (see RADIOACTIVITY) involves an actual breakup of the atomic nucleus. This is a *nuclear reaction* and involves energy exchanges millions of times greater than those of the ordinary electron reactions. Enough energy is involved to make a hydrogen bomb explosion or, if the mass involved is large enough, a sun.

Occultation

THE VARIOUS stars are so far from us that they seem to keep the same relative positions year after year. Not so the sun, moon, and planets, all of which appear to move in relation to the stars (see PLANET). It consequently happens that occasionally the moon, for instance, will pass in front of a star or planet, which will then remain hidden behind the moon for a period of time.

The Latin word for "to hide" is "occulare," so anything that is "hidden" is *occult*. This term is frequently applied to supposedly mysterious arts hidden to all but a few experts. The star or planet that is hidden by the moon is also occult and the hiding process is *occultation*.

Occultations involving the moon have been most useful to astronomers. For one thing, at the moment of occultation, the precise position of the moon in the sky is known so that occultations can be used to calculate the moon's orbit with considerable exactness. Secondly, the sudden disappearance of an occulted star, without preliminary dimming, is one of the proofs that the moon has no atmosphere to speak of. If there were an atmosphere, the star would first dim, as it shone through that atmosphere prior to occultation.

The most dramatic use of occultation, however, involved the planet Jupiter. The four large moons of Jupiter, as they circle their planet, pass behind it and are each occulted once a revolution. In 1676, the Danish astronomer Olaus Roemer had been timing those occultations and discovered an odd fact. When the earth was moving away from Jupiter in the course of their orbital motions, the time of occultation lagged and was later than it should have been. When the earth was moving toward Jupiter, the reverse happened; occultations grew more and more premature.

He deduced from this that it took light a finite time to travel through space. The distance between the earth and Jupiter could vary by about 200,000,000 miles and, apparently it took light about 16 seconds to make that distance.

It was fifty years before Roemer's work received much attention but today the speed of light (186,272 miles per second) is one of the fundamental constants of theoretical physics.

Octave

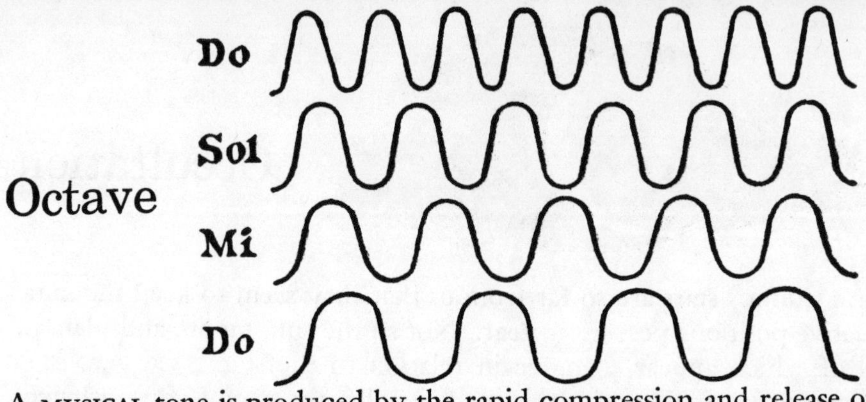

A MUSICAL tone is produced by the rapid compression and release of air by a vibrating object. This compression and release spreads out as longitudinal waves. The pitch of the tone depends upon the number of times per second (frequency) that the air is compressed and released. For instance, middle C ("do") on the piano has a frequency of 264.

Combinations of musical tones sound pleasant to the ear when the individual frequencies are in simple proportions. For instance, the frequency of *do* is 264, of *mi* 330, and of *sol* 396. These frequencies are 4×66, 5×66 and 6×66, so that the three notes form a pleasant sequence and sound well if struck together as a *chord* (from the Greek "chorde," meaning the string of a musical instrument).

The combination *fa, la* and *do* — and also *sol, ti* and *re* — form similar 4,5,6 ratios and sound pleasant chords. In fact the musical scale *do, re, mi, fa, sol, la, ti, do* was chosen just to allow a large number of pleasant combinations. The frequencies are: 264, 297, 330, 352, 396, 440, 495, and 528. There is nothing sacred about this; other note combinations are possible. However, western ears are used to this one, and the use of other combinations by the Chinese or Arabs makes their music sound odd to our ears.

Notice that the second, or higher, *do* has a frequency of 528 which is just twice that of the first *do* (264). By doubling all frequencies, we can build a new scale starting with the 528 *do* and ending with a still higher 1056 *do*. Or we can work downward from the 264 *do* to a still lower 128 *do*.

A new series is started with every eighth note, therefore. The Latin word for "eighth" is "octavus," so an individual series from *do* to *do* is an *octave*. Furthermore, any note with a frequency just twice that of a second note is said to be an octave higher. If it is four times the frequency, it is two octaves higher, and so on. (This notion has been extended to other forms of waves, too, as, for instance, light waves.)

Oil

When we are not certain of the exact nature of substances, we do what we can to differentiate them on the basis of outward appearance. For instance, tissues contain substances that are greasy in feel and do not dissolve in water. These substances may be solid or liquid. The difference is not important, really, since the solid can be melted and the liquid congealed.

Nevertheless, solid and liquid was all the difference men had to go on originally so two names are generally given. In English, the solid is *fat*, the liquid, *oil*. The word *fat* comes from the Anglo-Saxon, but *oil* is from the Latin "oleum," meaning "olive oil," which is related to the Greek "elaia," meaning the olive tree. (Olive oil was the chief oil for cooking and washing the body — no soap in those days — in the ancient Mediterranean world.)

The word *oil* has passed on to anything that feels oily. For instance, there are materials that can be obtained from plants that seem to possess the very essence of the plant in concentrated form. Roses will yield a small quantity of fluid containing all the fragrance of roses. Similarly, fluids can be obtained that have the fragrance or flavor of jasmine, cloves, vanilla, wintergreen, and so on. These fluids are oily to the touch and won't dissolve in water, so they are called *essential oils* (from *essence*).

Again, there is an oily liquid that is obtained from underground which is called *petroleum*. The "petr-" prefix comes from the Latin word "petra" (rock) so that the word means "rock oil." One of the substances obtained from petroleum is called *mineral oil*.

Very strong sulfuric acid has an oily appearance (it may have an oily feel, but no one would dare touch it) and is called *oleum*, though anything less like olive oil in other ways would be hard to imagine.

In contrast to all this, the Greek word for a solid fat is "stear." This enters chemistry at many points. For instance, one of the more common substances derived from solid fats is *stearic acid* and the word for soapstone is *steatite*.

Organism

The greek word "ergon" means "work" (see ENERGY) and from it the Greeks derived the word "organon" for any instrument that did work. This has come down to us as *organ* which we apply chiefly to a type of musical instrument, but the more general meaning still shows up. A branch of the government, such as the courts or the legislature, which does a particular type of work is an organ of government and a newspaper is an organ of the press.

The most important use of the term outside of music, however, is for any structure in a living creature that does a specific job of work. The heart, liver, lungs, skin are all *organs*. And since most living creatures are a collection of organs working together, they are, generally speaking, organisms. The word *organism* has lost its original significance and come to mean any living creature; thus a virus which may consist of a single large molecule only, and which cannot conceivably have organs in the usual sense, is still a *microorganism*, from the Greek "mikros," meaning "small" (see MICROBE).

By 1800, it seemed quite obvious that there were considerable differences between chemicals found only in living tissue (or in material that had once been part of living tissue) and chemicals found in the nonliving world all about. For one thing, chemists in the laboratory could not make any of the chemicals of living tissue unless they began with living tissue or the dead remnants thereof. In 1807, the Swedish chemist Jöns Jakob Berzelius divided all chemicals into two groups. Those from organisms, living or dead, he called *organic;* the rest he called *inorganic* (the Latin prefix "in-" means "not").

But then, in 1828, the German chemist Friedrich Wöhler upset things by preparing an organic chemical without any use of tissue, living or dead. Nevertheless the distinction remains useful and has been retained, but today *organic* refers to any compound with a molecule containing carbon atoms, while *inorganic* refers to any compound with a molecule that does not.

Ounce

It was not until modern times that the world has seen carefully standardized weights and measures on a universal scale. Until the nineteenth century, every country and region had its own system of weighing and measuring things. Naturally, this led to confusion. Under such conditions, merchants would want to use only those systems that were known to be reliable.

The city of Troyes in northeastern France was a flourishing city in the Middle Ages, noted for its fairs. To keep its business going, it regulated its weights and measures strictly; its unit of weight, the *pound* (from the Latin "pondus" meaning "a weight"), was widely adopted for valuable goods like gold, silver, jewels, and drugs, where a small inaccuracy could make a large money difference.

The pound used even today in weighing such materials is still called the troy pound, or sometimes the apothecaries' pound (because it is used in weighing drugs). The troy pound is divided into twelve troy ounces. *Ounce* comes from the Latin "uncia," meaning "a twelfth." (*Inch* is a slightly more distorted version of the same word; it means a twelfth of a foot.)

Potatoes, coal, and other substances which are relatively cheap and bulky, and which are usually handled in considerable weights, are measured by the avoirdupois pound. *Avoirdupois* is just the French phrase "avoir du pois," meaning "to have weight."

The avoirdupois pound is made up of sixteen avoirdupois ounces, and in this case, the word *ounce* is a misnomer, obviously. Nor are the troy ounce and the avoirdupois ounce equal. Both are divided into *grains* (a reminder of the time when grains of wheat or other cereals were used to weigh small quantities on hand scales, so that a tiny sliver of gold would weigh so many grains, you see). The troy ounce is equal to 480 grains, while the avoirdupois ounce is equal to only 437½ grains. The troy pound is equal to 5760 grains; the avoirdupois pound, to 7000 grains exactly.

Oxygen

The German chemist G. E. Stahl, about 1700, developed a theory to explain why some substances burned or rusted when heated. He supposed such substances contained *phlogiston* (from a Greek word "phlogistos," meaning "inflammable").

When wood was heated, for instance, it lost its phlogiston to the air and changed to ash. If the air supply were limited, then the air, after a while, would have all the phlogiston it could hold and burning would stop.

In 1774, the British clergyman and chemist Joseph Priestley studied a brick-red powder, today called mercuric oxide. He found this gave off a remarkable gas when heated. Substances burned in it more readily than in air itself. A glowing wooden splinter burst into brilliant flame if inserted into this new gas.

Well, if air that was full of phlogiston could not support combustion, then a substance that supported combustion so well should be completely empty of phlogiston. Priestley therefore called it *dephlogisticated air*.

The next year, the French chemist Antoine-Laurent Lavoisier showed that burning was simply the result of combining chemically with this new gas, which occurred naturally in air. The phlogiston theory fell apart and Lavoisier earned his title, "Father of Modern Chemistry."

But the great chemist was human and also made mistakes. He thought the new substance was found in all acids. (It isn't, but hydrogen is.) On the basis of this mistaken notion, he named the gas "oxygine" (*oxygen* in English), from the Greek word "oxys," meaning "sharp" and the suffix "-genes," meaning "born." Oxygen is then "that of which a sharp taste (or sourness) is born," and the process of combustion came to be called *oxidation*.

The Germans went along with this mistake but used their own language for the name, calling oxygen *Sauerstoff* ("sour substance").

Pachyderm

Most animals, if they are common, have names stretching back to antiquity, with their ultimate derivations uncertain (cat, mouse). Or, if they are first met with by explorers in foreign lands, the native names are sometimes adopted (chimpanzee, opossum). Some animal names, however, have origins that are worthier of discussion.

The *elephant*, for instance, derives its name from the Greek "elephas." This may in turn derive from the Phoenician "aleph" (ox). The elephant, after all, when brought to west Asia from India must have impressed the Asians with its sheer size. The largest animal with which they were acquainted was the ox, so this was called a kind of ox. (The Romans, when they first came across elephants, while fighting the Greek general Pyrrhus, in Lucania in southern Italy, called them "Lucanian oxen.")

The *hippopotamus* has a name that is pure Greek, coming from "hippos" (horse) and "potamos" (river). It is a "river horse." The hippopotamus does frequent certain African rivers and it is a large animal, so the Greeks naturally named it after another, and more familiar, large animal, the horse (although the hippopotamus looks about as much like a horse as the elephant does like an ox).

The elephant and hippopotamus were once included among the *Pachydermata* (from the Greek "pachys," meaning "thick," and "derma," meaning "skin"; i.e., the "thick-skinned" animals), all of which were hoofed animals that did not chew the cud, so that horses and pigs were included. This grouping has been abandoned but elephants are still sometimes called *pachyderms.*

The most conspicuous and unique feature of the elephant is its trunk, which is, in Greek, called "proboskis," (*proboscis*, in English) so that the elephant belongs to the order *Proboscidea.* The word *proboscis* comes from the Greek, "pro-" (before) and "boskesthai" (to graze). An animal with a proboscis can graze a long way before itself, and that is a good description of what an elephant can do.

Parabola

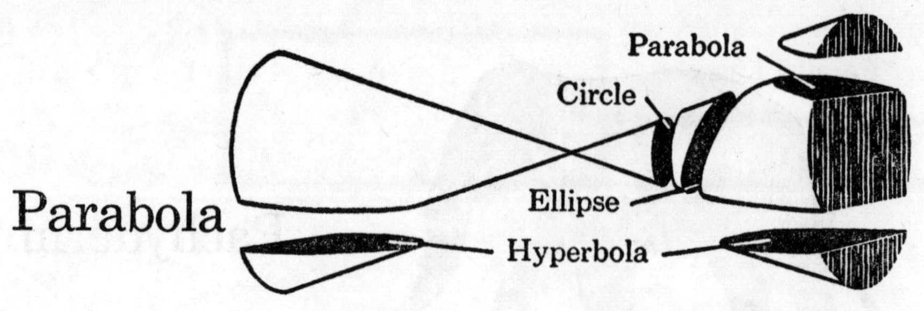

A CONE (from the Greek "konos") is a solid figure that looks like a dunce cap. If you were to imagine a cone cut by some kind of slicing machine in different ways, the cut edges of the cone would make certain interesting mathematical figures. For instance, if a cone were to be sliced straight across, so that all points on the sliced edge were equally distant from the point of the cone, the edge would form a circle. If the slice were made obliquely so that one side was farther from the point than the other, the edge would form an ellipse (see ELLIPSE). Circle and ellipse are examples of *conic sections*, the word *section* coming from the Latin "secare" (to cut).

In the case of both circle and ellipse, the slice goes completely through the cone from one side to another. However, if you imagine the slice starting at one side of the cone in a direction that was parallel (see PARALLEL) to the other side, the cut would continue forever, without ever reaching the other side (if we assume the cone to be of infinite size).

The edge of the cone formed by such a section would be an open curve. Unlike the circle and ellipse, a point traveling along it would never return to its starting place. The Greek geometer Appolonius of Perga (about 220 B.C.) called such a curve a *parabola*, from the Greek "para" (beside) and "ballein" (to throw). He named it so from the mathematical properties of the curve, but we can see the reasoning most easily by supposing that the imaginary knife making the section was thrown exactly and evenly beside (i.e., parallel to) the far edge of the cone.

It is also possible to make a section of a cone by cutting it in such a way that the cut actually recedes from the far edge of the cone. The curve thus formed is also open and Appolonius called it a *hyperbola.* The Greek prefix "hyper-" means "over" or "beyond." It is as though the knife were thrown with the intention of making a parallel cut but had gone "beyond" the mark. In language we still call an extravagant exaggeration by the Greek form of the word, *hyperbole.*

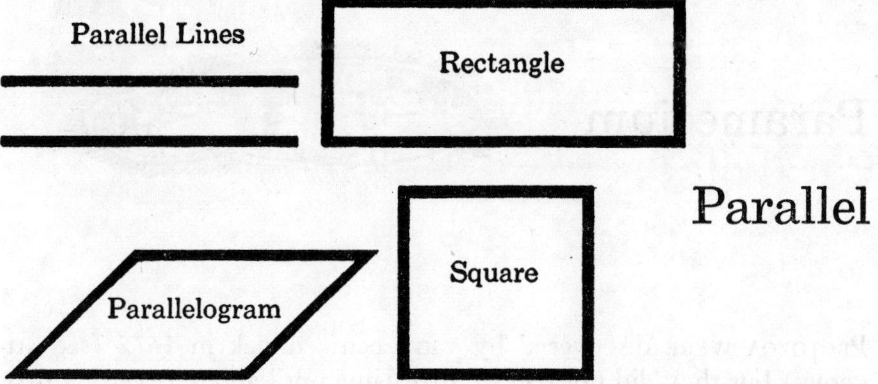

Parallel

THE FIRST line was a textile cord, rather than a mark on paper. *Line* comes from the Latin "linea" which probably comes from the old name for "flax" (hence, our word *linen* for what may be man's oldest textile). Since the word *straight* is but another version of "stretch" (see ANGLE), the phrase *straight line* is related, in derivation and sound, to "stretched linen," and it was a stretched linen cord that civilized man may have first used as a straight line in making land measurements.

Two straight lines running side by side indefinitely, neither approaching nor receding, are *parallel,* from the Greek words "para" (beside) and "allelon" (of one another). The Greek word for "line" is "gramme," so a four-sided figure made up of two intersecting pairs of parallel lines is a *parallelogram.*

If the two pairs of parallel lines meet perpendicularly so that all four angles are right angles (see HYPOTENUSE), the figure is a rectangle. The prefix "rect-" comes from the Latin "rectus" (right); thus *rectangle* and *right angle* are different versions of the same word, even though the former refers to a closed figure, the latter to a single angle.

In general, any four-sided figure, including the parallelogram and the rectangle, is a *quadrilateral,* from the Latin words "quattuor" (four) and "latus" (side). It is "four-sided." The number four pops up again in the quadrilateral with all angles right angles and all sides equal. The Latin prefix "ex-" means "out" and "quadri-" is the common Latin combining form derived from "quattuor" (four). To draw a figure out of four lines (and making it, most naturally, the simplest and most regular possible) is to make it "ex quadre." This becomes "esquarre" in Old French and *square* in English.

Paramecium

PROTOZOA WERE discovered by van Leeuwenhoek in 1675 (see MICROBE) but they did not receive that name until about 1818. At first, the one-celled creatures were simply called *animalcules*, from the Latin diminutive "animalculum" (a little animal).

These animalcules could be easily obtained for study. It was only necessary to take some vegetable matter, steep it in water, and expose it to air. From those animalcules that might have been on the vegetable matter or in the water to begin with, or from those that were blown in, the numbers increased. Something soaking in water is called an *infusion* from the Latin "in-" (in) and "fundere" (to pour); water is "poured in" to make an infusion, in other words. Because of their appearance in infusions, the animalcules were given the general name of *Infusoria* about 1763.

Today, though, Infusoria is a name restricted only to the most highly developed group of the protozoa, to little creatures that almost ape the features of more complex organisms. There is a definite spot on the cell membrane where food is taken in and another where wastes are eliminated. The cells have definite shape and move about by the rapid and coordinated movement of tiny hairlike filaments all over the cell surface. These filaments are *cilia*, which is a Latin word meaning "eyelashes." Even here, though, the name *Ciliophora* is replacing *Infusoria.* Since the Greek "pherein" means "to bear," *Ciliophora* means "cilia-bearing."

The most familiar of the ciliophores is the paramecium. The front of the paramecium cell is pointed and the back is rounded while there is a constriction in the middle. In outline, it has the shape of a slipper, so much so that a common name for it is "slipper animalcule." *Paramecium* itself is a much less descriptive name since it comes from the Greek "paramekes" which means simply "oblong."

Some nonciliophores move by a similar method. They possess one or two long cilia that whip about and drive the cells along. These long cilia are called *flagella*, a Latin word meaning "whips."

Parsec

You will observe that if, while looking at an object, you move your head, the object seems to move in the opposite direction as compared with the distant horizon. Furthermore, a distant object seems to move less than does a nearby object. For that reason, a nearby tree might hide a particular house in the distance at first, and a different house after you have moved.

This apparent motion of near objects in comparison with far objects as the viewer moves is called *parallax* from the Greek words "para" (beside) and "allassein" (make otherwise). In other words, with the motion of your head, that which the near object was beside is made otherwise.

The most dramatic use made of the phenomenon of parallax is in the measurement of the distance of stars. As the earth moves about the sun, the stars seem to move in tiny ellipses in the opposite direction. The stars are so far away that these ellipses are very small indeed.

The semi-major axis of this circle is the *stellar parallax* (the Latin word "stella" means "star") and this is always less than a second. (The circuit of the sky is divided into 360 degrees, each degree into 60 minutes, and each minute into 60 seconds.) The nearest star, Alpha Centauri, for instance, has a parallax of about three fourths of a second. Such a parallax, considering the position shift of the earth, corresponds to a distance of 25,000,000,000,000 miles.

This is an inconveniently large number to handle. A large unit is needed to measure astronomical distances. Light travels 186,272 miles per second, or 5,880,000,000,000 miles in a year, so the latter distance is known as a *light-year*. Alpha Centauri is, therefore, 4.25 light-years away from us.

Another way of measuring large distances is to take as a unit that distance at which a star will have a parallax of exactly one second. This distance is 3.26 light-years and such a distance is said to be one *parsec*, a made-up word combining the first syllables of *parallax* and *second*.

Penicillin

PEOPLE AND animals who die of disease are buried underground and yet the soil remains fairly free of disease germs. The bacteria and other microscopic organisms that live in soil destroy them.

In 1929, the English physician Alexander Fleming became sharply aware of this when he noticed that spores of bread mold had gotten into a culture of disease germs he was growing, and that around each spore was a clear area where no germs grew. Fleming decided there was a chemical in the mold which stopped the germs. Since the scientific name of the mold was *Penicillium notatum* he named the chemical *penicillin.* (He shared the 1945 Nobel Prize in medicine for this.)

When World War II began, an Anglo-American research effort was put forth, penicillin was isolated, its structure determined and large-scale production begun. Since the war, penicillin and substances like it have largely replaced the sulfa drugs (see SULFANILAMIDE) and a number of diseases and infections have suddenly been brought under control.

Beginning in 1940, the Russian-born American microbiologist Selman A. Waksman produced a number of bacteria-killing compounds. From molds known as *Streptomyces* (from the Greek "streptos" (twisted) and "mykes" (fungus) — a fungus with twisted threadlike structures) he isolated one he called *streptomycin.* In 1942, he proposed the name *antibiotic* for such compounds, from the Greek "anti-" (against) and "bios" (life); that is, they acted "against (germ) life." (Waksman received the Nobel Prize in medicine in 1952).

Another group of molds, the *Actinomyces* (from the Greek "aktis" (ray) — fungi with radiating threadlike structures) includes *aureomycin* (from the Latin "aurum" (gold) because of its golden color), *terramycin* (from the Latin "terra" (earth) because it came from molds that lived in the soil) and *achromycin* (from the Greek "achromos," meaning "without color").

These last consist of molecules made up of four circles of carbon atoms joined together and are now called the *tetracyclines,* from the Greek "tettares" (four) and "kyklos" (circle).

Perihelion

The ancient Greeks always insisted that heavenly bodies moved in orbits that were exact circles, because the circle was the perfect curve and all things heavenly were perfect. The very word *orbit* comes from the Latin word "orbis," meaning "circle."

In 1609, however, the German astronomer Johann Kepler finally showed that planets moved about the sun in ellipses, not in circles, and that the sun was not in the center but at one of the foci of the ellipse (see ELLIPSE).

Since the sun was out of center, it meant that at some points in their orbits the planets would be closer to the sun than at other points. The point at which the planet is nearest the sun is the *perihelion*, from the Greek words "peri," meaning "around," and "helios," meaning "sun." It is a poor word unless you take "around" as meaning "in the neighborhood of," in which case *perihelion* is the point at which the planet is "in the neighborhood of the sun." Similarly, the point at which a planet is farthest from the sun is the *aphelion*, the prefix coming from the Greek word "apo," meaning "from." At aphelion a planet is away from the sun.

The ellipse of the earth's orbit is of low eccentricity so that at perihelion, the earth is only 3 per cent closer to the sun than at aphelion.

Similar prefixes are used when an orbit is about some body other than the sun. For instance, the moon travels about the earth in an ellipse with the earth at one focus. The point of closest approach of the moon is *perigee*, the suffix "-gee" coming from the Greek word "ge," meaning earth. The point of the moon's greatest distance from the earth is *apogee*.

There are many known instances of two stars circling about their common center of gravity. Each travels in an ellipse with the center of gravity at one focus. The more massive star travels in the smaller ellipse. The point where the two stars approach most closely is the *periastron* (from the Greek word "astron," meaning "star") while the point of greatest distance is the *apastron*.

Perturbation

There is something innately disorderly about a crowd. Even if they are perfectly peaceable, the mere fact that some are heading in one direction and some in another makes for disorder. Or if they are all standing still, there are always some turning heads one way, some mopping their brows, some fidgeting. The great wonder about the Radio City Music Hall Rockettes, in fact, is that it is possible to take even as few as 48 people and make them do the same things at the same time.

The Latin word for "crowd" is "turba" and a number of words implying disorder contain that as a stem. Water which has been stirred up so that mud from the bottom is mixed with it is *turbid.* Water being whipped up by waves (or emotions being whipped up by passion) are *turbulent.* A person who is upset and in disorder is *disturbed* or *perturbed.*

Now the law of gravity can only be applied exactly when no more than two bodies are involved. For instance, the earth travels about the sun in a course that would be exactly determined if the earth and sun were the only objects in the universe. However, the moon pulls at earth a bit and so do Mars and Venus and, in theory, all other bodies in the universe.

Fortunately, the effects of these other bodies on the earth is minor compared to that of the sun. So earth's orbit can be calculated as though the other bodies did not exist and then their minor effects can be worked out by noting the manner in which earth departs from its calculated orbit.

These minor gravitational effects of astronomical bodies upon each other, which cause a departure from the stately order of travel according to the "two-body" calculation, and set up, so to speak, a certain amount of disorder, are called *perturbations.* These had their most dramatic moment in astronomical history when perturbations in the orbit of Uranus, which could not be explained by the effect of the other known planets, led to the suggestion by John C. Adams of England and Urbain J. J. Leverrier of France that an unknown outer planet lay beyond Uranus and could be found at that time in a certain spot of the heavens. Theory was verified by observation and, in 1846, Neptune was discovered.

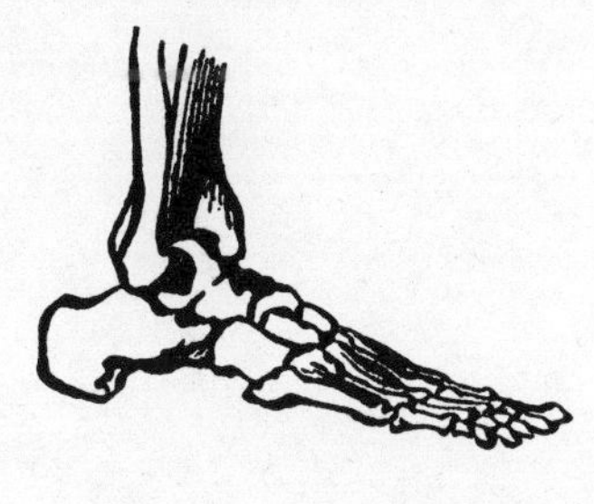

Phalanges

PERHAPS THE greatest military genius among the ancient Greeks was Epaminondas of Thebes. Until his time, Greeks had always lined up their infantry in straight lines several ranks deep and let them clash in battle. The Spartans, with the best-trained infantry, invariably won, if they were involved.

Epaminondas, however, set up the right flank of his army some fifty ranks deep. The center and the left flank, each very light, was set back so that the heavy flank struck first. Its sheer weight smashed through the opposing line, like a large log of wood swinging against a door, and completely disorganized the enemy. This tactic was first used at the battle of Leuctra in 371 B.C. where Epaminondas defeated the Spartans completely and ended their role as a first-class military power at once and permanently.

The Greek word for such a mass of infantry — "phalanx" — is of uncertain derivation, but it is also their word for "log" and perhaps the infantry mass was likened to a swinging log, as I have done in the paragraph above.

In any case, Philip II of Macedon (a semi-Greek borderland) was a prisoner of war at Thebes during the time of Epaminondas and later, as a monarch, remembered what he had seen. He adopted and improved the phalanx, lightened it and increased its maneuverability, supported it well with cavalry, and equipped each man with a long spear so that the phalanx bristled like a porcupine. With it, he conquered the Greeks, and his son, Alexander III (the Great), conquered the Persian Empire.

The most noteworthy thing about the phalanx was the close order maintained by the men; ranks and files kept together. This same close order is maintained by the bones in the fingers and toes. In each finger and toe are three small bones (two only in thumbs and big toes), one immediately behind the other. There are five sets of them, side by side, in each hand and foot (56 bones all together) and because of their close order they are known to anatomists as *phalanges,* a plural form of "phalanx."

Phobos

In the latter half of the 1800's, as far as was known, the earth had one moon; Mars (next out from the sun) had none; Jupiter (next out) had four, and Saturn had eight. To make a perfect sequence, Mars should have had two moons; then it would be: 1, 2, 4, 8.

Of course, astronomers didn't take that sort of number-juggling seriously, though some people did. In *Gulliver's Travels*, a satire written in 1726 by Jonathan Swift, the inhabitants of a mythical land named Laputa were supposed to have discovered two moons of Mars by means of superior telescopes, and Swift described the moons in considerable detail.

Then, in 1877, when Mars was at a point in its orbit close to earth, the American astronomer Asaph Hall decided to make a systematic search for any satellites Mars might have. He searched methodically, night after night, until it seemed certain there was nothing. He went home, discouraged, one day, having given up, when, as the story has it, Mrs. Hall persuaded him to go back and try one more time.

And, as often happens in fiction, but practically never in real life, that one more time was it. Hall spotted something. He had to wait out a siege of cloudy weather to check it, and when he did, there was not one satellite, but two. What's more their smallness, their closeness to the planet, their speed of revolution were all very unusual and yet all very close to what had been described by Swift. Certainly Swift's was probably the most inspired guess in all literature.

Naming the Martian satellites was easy. The Greeks had Ares (the Roman Mars) as god of war, attended by his two sons, Phobos (Greek for "fear") and Deimos (Greek for "terror"). Hall called the inner satellite of Mars *Phobos* and the outer *Deimos*, so that Mars in the sky, as well as in myth and in reality, is attended perpetually by Fear and Terror.

But alas, the sequence of moons was spoiled anyway. In 1898, a ninth moon of Saturn was discovered and in 1901, a fifth moon of Jupiter. As a matter of fact, twelve moons of Jupiter are now known.

Phosphorus

The planet Venus is closer to the sun than we are. As a result, when viewed from earth, it sometimes appears to the east of the sun, sometimes to the west; but in neither case does it get very far from the sun.

When Venus is east of the sun, it reaches the western horizon later, so that for a time after the sun sets, it remains glowing brightly in the western sky. It is then the *evening star,* which the Greeks called Hesperos ("evening"). In the morning, Venus reaches the eastern horizon later than the sun; by the time Venus rises, the sun is in the sky and Venus's light is drowned out.

When Venus is to the west of the sun, it reaches the western horizon sooner than the sun and by sunset it is gone. However, in the morning, Venus reaches the eastern horizon sooner than does the sun and for a time before the sun rises it glows brightly in the eastern sky. It is then the *morning star,* which the Greeks called *Phosphoros.* This word comes from "phos" (light) and "phoros" (bearing). The morning star was the "bearer of light," for, once it appeared, the sun could not be far behind.

Since Venus cannot be both east and west of the sun simultaneously, the evening star and morning star can never appear on the same day. When the evening star is in the sky there is no morning star, and vice versa. This gradually dawned on the Greeks and they came to realize that the two stars were one planet, which they then named the star of Aphrodite, the goddess of love, called by the Romans Venus.

The word *Hesperos* (Latin, Hesperus) does not appear in English except in poetry, but *Phosphoros* has taken on a new and permanent lease of life. In about 1669, a German alchemist named Hennig Brand isolated a waxy white substance from urine, which glowed in the dark. (It combined with oxygen and the energy of combination liberated light, but Brand did not know that.) Because it was a "light-bearer," he called it *phosphorus,* and what was once the name of the morning star is now the name of a chemical element.

Phylum

ONE WAY of dividing the animal kingdom is to include those animals which, like ourselves, have a bony skeleton under the heading *vertebrates.* Although these skeletons may differ in detail from animal to animal, they all include a backbone. The individual bone of a backbone is a vertebra (see VERTEBRA) and hence the name *vertebrate.*

All other animals may then be dismissed as *invertebrates;* that is, having no internal bony skeleton.

However, this is not a logical system of classification. Among the invertebrates, there are groups of animals differing from one another as much as any of them differ from the vertebrates. Zoologists have therefore divided the animal kingdom into a number of groups, each called a *phylum* (from the Greek "phylon" meaning "tribe"). All the animals included within a phylum have the same general body plan, varying only in detail. Thus, the vertebrates make up a single phylum, because all have a bony skeleton, a front and back end, no more than four limbs, and so on.

The invertebrates, however, are divided into more than twenty phyla, each with its own body plan. Some are small groups including a few odd animals known only to zoologists. Others are rich collections of animals more numerous and various than those of the vertebrates.

For instance, insects, spiders, crabs, lobsters, and similar creatures are all classified in the phylum *Arthropoda,* from the Greek "arthron" (joint) and "pous" (foot). They all have jointed limbs and bodies, too. They are covered by a shell which is also jointed.

Snails, slugs, oysters, clams, and squids, on the other hand, are members of the phylum Mollusca. These also have shells (or sometimes the remnants of one) but their shells have a different chemical composition from those of arthropods and are not jointed. Within the hard shell is a soft, gooey creature (have you ever eaten oysters?) and *Mollusca* comes from the Latin "molluscus," meaning "soft."

Pithecanthropus

THE GREEK word for man is "anthropos" and this shows up at once in the name of that science which concerns itself with the human species, *anthropology*. Again, those apes which most closely resemble man are called the *anthropoid apes* (the "manlike" apes).

However, the most dramatic use of the word occurs in connection with certain extinct creatures that are not exactly men and yet are closer to the human than is any living ape. On the basis of the different species of such "ape men" that have been found, anthropologists have tried to trace the possible route by which "true man" has developed.

These antique specimens are often called, familiarly, after the regions in which the remains were discovered; for instance, we have Peking man, Java man, Heidelberg man, Rhodesian man, and so on. However, anthropologists have tried to name them, in Latin, by genus and species, as they have done with other creatures, both living and extinct.

For instance, Peking man, the most ancient of the ape men, has the scientific name of *Sinanthropus pekinensis*. The prefix "Sin-" is a version of "Chin-" (as in "Sino-Japanese war," for instance), so the name means "China man of Peking."

Java man, one of the first of the ape men to be found, was named as long ago as 1891, and given the title of *Pithecanthropus erectus*. Since the Greek "pithekos" means "ape," the name actually means "ape man," the "ape man who stands erect."

True man also has a Latin name. He belongs to the genus *Homo* (the Latin word for "man"). The Neanderthal man is a primitive variety of true man and is called *Homo neanderthalensis*. (Skeletons were found most notably in the valley — German, *Thal* — of the Neander River, a tributary of the Rhine.) We ourselves are *Homo sapiens* (Latin for "Man, the Wise"). Perhaps, in comparison with other living creatures, this is an appropriate name, but sometimes, alas, it seems as though the slang abbreviation "Homo sap" (Man, the Fool) may be more appropriate.

Pituitary

Before the nineteenth century, no one really understood the function of the brain. The best the ancient Greeks could do was to suppose that it acted as an air conditioner, cooling the overheated blood, or that it produced the mucus or phlegm which is the thickish fluid that coats the inner membranes of nose and throat and is most noticeable when we have a cold. ("Mucus" is a Latin word and "phlegma" a Greek one. There is an Anglo-Saxon word, too, which, however, is not used in polite society.)

Presumably, the Latins didn't like their own word, either, and used "pituita" instead, which they may have felt to be politer. It may perhaps be related to the Greek "ptyein" (to spit) and an unpleasant word always seems nicer if it is in a foreign language. In any case, the phlegm-producing function of the brain was centered in a small organ (they thought) attached to the bottom of the brain by a thin stalk, and that organ was therefore called the *pituitary*.

It amounts to a kind of Cinderella story, then, to point out that this small organ with the rather repellent name has turned out to be the master gland of the body. It produces a series of hormones which control the action of other glands. It has a less ludicrous name, the *hypophysis* (a Greek word meaning "undergrowth"; it "grows under" the brain), introduced about a hundred years ago by the American anatomist Burt Green Wilder, but it is almost always called the pituitary.

Many of the pituitary hormones have the stem "troph" in their names from the Greek "trephein" (to cause to grow) since they seem to feed other glands and improve the yield of other hormones. This stem has been corrupted to "trop," however. Thus, one of the hormones is *adrenocorticotropic hormone* (which "feeds the adrenal cortex" and improves the yield of cortical hormones; see CORTISONE). This name has been abbreviated to *ACTH* and, in this guise, the hormone has gained considerable newspaper fame as one of modern medicine's "miracle drugs."

Planet

EVEN BEFORE the dawn of written history, men watching the stars at night must have noticed that they never changed their pattern. A group of stars forming a lopsided W kept that shape night after night, lifetime after lifetime. These were the *fixed stars*. As a group, the stars moved about a point in the northern skies. Those far enough away from that central point rose and set as did the sun and moon.

Each night, the entire star-pattern shifted a little. Each star rose four minutes earlier and set four minutes earlier than the night before, so that stars in the west gradually disappeared behind the horizon while other stars appeared in the east. After a full year, the circle was complete and the original pattern was again in the night sky.

There were, however, five starlike objects, as bright or brighter than the brightest stars, that did not stick to the pattern. One of these objects might lie midway between two stars on a certain night, for instance. The next night, it would have shifted its position; the next night, it would have shifted farther, and so on. Three of these starlike objects (we call them Mars, Jupiter, and Saturn) made a complete circuit of the sky, by a rather complicated route. The other two (Mercury and Venus) did not move more than a certain distance from the sun.

These objects, in other words, were not fixed stars, but "unfixed" ones. They "wandered" among the stars. The Greek word for a wanderer is "planetes" and so the Greeks called these bright objects by that name and it has come down to us as *planets*.

In ancient and medieval times, the sun and moon were numbered among the planets, because they wandered among the stars, too. By the seventeenth century, however, the fact was accepted that the sun was the center of the solar system, and a planet then became any astronomical body that revolved about the sun. The sun itself was no longer a planet, then, but the earth became one. The moon, too, is not a planet since it revolves about the earth, primarily, and only secondarily about the sun.

Plankton

LIKE THE animal kingdom, the plant kingdom is systematically divided into groups and subgroups. There is first a grand division into two subkingdoms, the more primitive of which is *Thallophyta.* These include all the one-celled plants, plus related many-celled plants in which the individual cells are comparatively little specialized.

The largest of these, and an example sure to be known even to people without microscopes, are the seaweeds. These consist of mere undifferentiated shoots (in Greek, "thallos") and lack roots, leaves, or true stems. The Greek "phyton" means "plant," so *Thallophyta* means "shoot-plants."

The Thallophyta are divided into a number of phyla (see PHYLUM) which in turn fall into two groups, those which include plants with chlorophyll (see CHLOROPHYLL) and those which include plants without. The former are called *algae* (singular, *alga*) which is the Latin word for seaweed, an alga that can be seen with the naked eye. The latter are *fungi* (singular, *fungus*), which is the Latin word for mushroom, a fungus that can be seen with the naked eye. (Bacteria are included among the fungi, usually.)

The plant life of the oceans (which makes up about 85 per cent of all the greenery on earth) belongs entirely to the Thallophyta. The algae of the oceans manufacture their food with the aid of sunlight and so must exist only in the topmost layers of the ocean where light can penetrate. There they float, drifting with the current and serving as food, directly or indirectly, for all the animal life of the sea. These algae are *phytoplankton.* The Greek "planktos" means "wandering," so phytoplankton are "plant-wanderers" because they must drift ("wander") with the ocean currents. Small animal life that also drifts in the currents makes up the *zooplankton.* The Greek "zoon" means "animal," so zooplankton are "animal-wanderers." Together, phytoplankton and zooplankton are referred to simply as *plankton.*

There are also animals in the surface waters which are independent of current and swim as they please. These are *nekton,* from the Greek "nektos" (swimming).

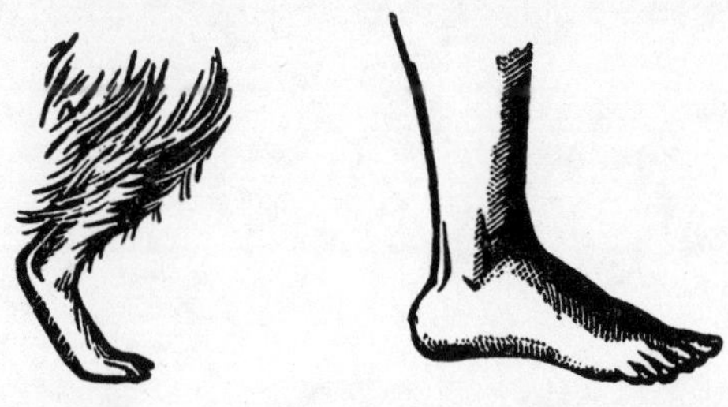

Plantigrade

The lower surface of the foot is the *sole,* a word which comes from the Latin "solea" (sandal); the sole is the part of the body covered by a sandal. Man walks on the entire sole of his foot, putting the foot down on toes and heels with each step. We are therefore *plantigrade,* from the Latin "planta" (sole) and "gradior" (to walk). We are "sole-walkers." Actually, though, only a minority of mammals walk in this way; we, the apes, bears, and raccoons are the chief examples.

Most mammals walk on their toes only. The heels are lifted high and in normal walking never touch the ground. Walking on toes has the advantage of adding the length of the foot to that of the leg. The animal is higher and can see farther; with longer legs, it can also run faster. These mammals are *digitigrade.* The Latin "digitus" means "toe," so they are "toe-walkers."

The extreme occurs in those mammals that lift up to the very toe-tips, which are sheathed in enlarged nails called hoofs (see UNGULATE). These are mostly animals which depend on running for safety (though the fastest land-runner, as it happens, is the cheetah, a large cat and certainly without hoofs).

The hoofed mammals are divided into two orders, depending on the number of hoofs. Those with two hoofs on each leg (cattle, sheep, goats, swine, deer, etc.) belong to the *Artiodactyla,* from the Greek "artios" (even) and "daktylos" (toe); they are the "even-toes." Those with an odd number of toes (horses, donkeys, zebras, with one hoof per leg; rhinoceroses and tapirs with three) belong to the *Perissodactyla,* from the Greek "perissos" (odd): the "odd-toes."

The Greek words for "odd" and "even" show signs of having originated from the necessity of sharing things. An even number of anything can be shared exactly and the Greek "arti" means "exactly"; hence "artios" is "an even number." An odd number when shared always result in one item left over. The Greek "peri" means "over," hence "perissos" for an odd number.

Pleistocene

THE MAIN eras of the earth's history (see FOSSIL) were divided into subgroups with names often derived from the places where the rock strata belonging to a particular period of time were first studied. For instance, rock strata belonging to the Paleozoic era were mainly studied in Wales, so the first three suberas were named the *Cambrian, Ordovician,* and *Silurian* periods after Cambria, the old Latin name for Wales, and after two of the Celtic tribes that had lived there in pre-Roman times, the Ordovices and the Silures. The Cambrian period is the earliest from which definite fossils are obtained. Though there are dim evidences of life before that, people who are interested in fossils generally lump all the periods before that as pre-Cambrian and let it go at that.

The next division of the Paleozoic era is the *Devonian,* from Devon, an English county just across the Bristol Channel from Wales. This is followed by the *Carboniferous* period. This breaks the place-name habit since it is derived from the Latin "carbo" (coal) and "ferre" (to carry). It was the era during which our modern coal beds were laid down, so it "carries the coal." The Carboniferous is, however, further subdivided into the Mississippian, Pennsylvanian, and *Permian* periods. The first two of these have obvious derivations; the last, as a change of pace, is named after the region of Perm in the Ural Mountains of the Soviet Union.

The most recent of the periods in time is the Cenozoic which is divided on a different plan into the *Eocene, Oligocene, Miocene, Pliocene, Pleistocene,* and *Holocene* periods. The suffix "-cene" comes from the Greek "kainos" (new) while the prefixes are derived from the Greek "eos" (dawn), "oligos" (few), "meion" (less), "pleion" (more), "pleistos" (most), and "holos" (whole). The periods, in other words, are: (1) Eocene, the dawn of the new; (2) Oligocene, a few bits of the new; (3) Miocene, less than half of the new; (4) Pliocene, more than half of the new; (5) Pleistocene, most of the new; and (6) Holocene, the whole of the new.

Polymer

LIVING TISSUE builds up its giant molecules by joining together strings of smaller molecules (see NEOPRENE and MONOSACCHARIDE). Sometimes, the giant molecules have properties that are startlingly different from the units of which they are composed. For instance, sugars are soft, powdery solids, but they can join together to form the fibrous compounds in wood, which are strong enough to build houses out of.

The chemist has tried to design chemicals with new and useful properties by imitating this process, beginning with some simple compound and encouraging it to combine into long-string giant molecules. The original unit molecule is the *monomer*, from the Greek "monos" (one) and "meros" (part); it is the "one part" with which you begin. The final giant molecule is a *polymer*, from the Greek "polys" (many); it is made up of "many parts." The process of joining up the original units is *polymerization.*

The best-known products of such laboratory polymerizations are the various *plastics*. The first product of a polymerization is a soft solid that can be molded into any shape desired. The Greek "plassein" means "to mold," and "plastikos" means "fit for molding." Sometimes the plastic, after cooling, can be heated and molded a second time, or any number of times, into a new shape. This is a *thermoplastic*, from the Greek "thermos" (heat). Other plastics, once cooled, set permanently into shape and cannot be molded thereafter. These are the *thermosetting plastics.*

Some familiar plastics are named after the monomer. For instance, one well-known plastic is built up out of molecules of a gas called *ethylene* (chemically related to the carbon chains in ether — see ETHER — whence its name) and is therefore called *polyethylene* ("many ethylenes"). Another monomer is *styrene* (a liquid obtained from the resin of a tree called Styrax or Storax, whence its name) and a plastic built up out of that is *polystyrene.*

Porphyrin

THE MOST noticeable thing about blood, to the naked eye, is its color and this crops up in the names of things that have nothing to do with blood. For instance, the most common iron ore is a brownish red mineral which, because of its color, received the name *hematite* (from the Greek "haima," meaning "blood" and "haimatites," "bloodlines") as early as Greek times.

As it happens, there is iron in the compound that gives blood its color, though it isn't the iron that's responsible. The red compound in blood contains a complicated ring of carbon and nitrogen atoms which is red-purple in color all by itself. It is called the *porphyrin* ring, from the Greek "porphyra," the name of the fish from which purple dye was obtained. (There are also red-purple minerals which have nothing to do with the porphyrin ring which are called *porphyry* because of the color.)

The particular porphyrin compound found in the red substance of blood is *protoporphyrin IX*. The "proto-" prefix comes from the Greek "protos" (first) because this particular variety of porphyrin is first in importance in the body. The reason for the IX is more complicated.

By making minor adjustments in the outer atoms of the protoporphyrin molecule, it is possible to imagine fifteen different compounds, all protoporphyrin varieties, each of which might qualify as the red compound of blood. Only one could really be that compound. The German physician and biochemist Hans Fischer numbered these compounds from one to fifteen and he and his students set about synthesizing each one in the laboratory. Each, as it was prepared, was compared with the natural product obtained from blood. It turned out that the particular protoporphyrin which Fischer had happened to number "nine" was the natural one, hence protoporphyrin IX. In 1930, Hans Fischer received the Nobel Prize in chemistry for his work on blood compounds.

The protoporphyrin IX exists in blood in combination with iron and the combination was first named *hematin*, but is now generally called *heme*, so the Greek word for blood ends by being given to the compound responsible for the color.

Potassium

SOAP WAS not known to ancient peoples. The Greeks and Romans used oil as a cleaning agent. This sounds odd to us now, but oil will dissolve grease and help remove grime. Naturally, any substance that would help the oil do this would be much in demand. The ancients showed their desperation in that they sometimes added sand or other gritty material to the oil, for the sake of the scouring action, but that has obvious disadvantages.

A more suitable additive was eventually found in the ash of certain woods. This ash would be stirred in water, which would dissolve out some of the substances in the ash. The water, with the substances dissolved in it, would be poured off into a large pot. The water would then be boiled off and the dry residue heated strongly. The powdery material that resulted was called (in English) *potash*. The two syllables run together in pronunciation so that most people don't notice that it simply means "pot ash." The Arabs, who were the great chemists of the Middle Ages, called the same material "alquili" meaning "the plant ash."

When oil was heated with potash, a kind of soap was formed, so that a new and better cleaning material was developed.

The British chemist Sir Humphry Davy isolated a hitherto unknown metal, in 1807, and, because it occurred in potash, he gave it the Latin-sounding name *potassium*. The Germans, oddly enough, gave it a Latin-sounding name derived from the same substance, but they used the Arabic word and called the metal *Kalium*. For that reason, the chemical symbol for the metal is K, even in those countries that call it potassium.

Potash is one of a group of substances that share properties opposite to those possessed by acids. Such opposite-to-acid compounds are named *alkalis*, a word which obviously descends from the Arabic word for "potash."

Power

UNTIL THE middle of the eighteenth century, work was done by the muscles of humans and animals or by such natural forces as the wind, which mankind found ready to hand. Periodically, men would try to do something with the force generated by boiling water as it expands into steam. The first really useful steam engine (see ENGINE) was constructed by the Scottish inventor James Watt in 1765 and patented in 1769.

Watt's steam engine was first used to work the pumps draining water out of mines. Previously, they had been worked by men or, more usually, by horses. The dryness of the mine depended on how fast the water could be removed, in other words, on the rate at which the work of lifting the water could be performed. The rate of doing work is termed *power* (from the Latin word "posse," meaning "be able").

Watt measured the power that could be exerted by a horse. By means of a rope and pulley, he found that a particular horse could lift a weight of 150 pounds through a height of 221 feet in one minute. Power is measured by multiplying the weight lifted by the height through which it is lifted and dividing the product by the time. One *horsepower* is, therefore, $150 \times 221 \div 1$ or, in the round numbers which have come to be accepted, 33,000 foot-pounds per minute.

By measuring the power of an automobile or airplane engine in horsepower, we are still harking back to the days when Watt was curious to know how many horses could be replaced by his steam engine.

In electrical measurements, the power of a current is measured in *watts,* in honor of James Watt. The power can be obtained by multiplying the volts of electromotive force (see VOLT) by the amperes of current strength. Thus, an electric bulb with a current strength of half an ampere under an electromotive force of 120 volts has a power of 60 watts. One horsepower, by the way, is equal to 745.2 watts.

Prime Number

GREEK MATHEMATICIANS were fond of playing with numbers and began games with them that still exercise the brains of our own mathematicians.

For instance some *numbers* (from the Latin "numerus" meaning "number") can be divided evenly by smaller numbers. For instance, the number 12 can be divided by 2, 3, 4, or 6. Twelve divided by two is six; twelve divided by three is four; twelve divided by four is three; and twelve divided by six is two. Each of these numbers is a *factor* of twelve (from the Latin "facere," meaning "to make," since out of these smaller numbers you can "make" the larger one).

Of course, all numbers can be divided evenly by the number 1 (twelve divided by one is twelve), or by themselves (twelve divided by twelve is one). Such "universal factors" are of no interest and can be ignored.

If we concentrate on the other factors, though, it turns out that some numbers possess none. The numbers 2, 3, 5, 7, 11, 13, 17, 19, 23, 29, 31, and 37, for instance, have no factors other than one and themselves. (And numbers of any size can lack factors. There is no upper limit.) Such factorless numbers are *prime numbers* or just *primes,* from the Latin "primus" (first).

The reason for that is that other numbers, those *not* prime, can be broken up into prime factors. The number 12, for instance, can be expressed as 3 × 2 × 2. (The number 2 is the only even prime. All other even numbers can be divided by 2 and so have that as a factor and are not prime.) The number 15, is 5 × 3; 143 is 11 × 13; 370 is 37 × 5 × 2, and so on. Such numbers, with factors, are *composite numbers,* from the Latin "com-" (together) and "ponere" (past participle "positus," to place); they are built up by "placing together" smaller numbers. So the primes can be viewed, in a poetic way, as having existed first, while other numbers came into existence afterward only by the building up of primes.

Protein

THE FIRST person to distinguish the three main groups of substances in food was William Prout (see PROTON), in 1827. He named them the saccharine, oily, and albuminous substances.

The word *saccharine* comes from the Greek "sakchar" (sugar) and includes the various *sugars* ("sugar," one can see, is a distant relative of "sakchar") and the starches, which can be converted to sugar by acid. *Starch* comes from the Anglo-Saxon "stearc" (strong), since when added to a limp textile, it will make it stiff. Sugars, starches, and related compounds are today lumped together as *carbohydrates.* This name originated through the mistaken notion on the part of early chemists that carbohydrate molecules consisted of a string of carbon atoms to which water molecules were attached (the Greek "hydor" means "water"). As usual, mere wrongness did not prevent the name from sticking.

Prout's oily substances included both oils and fats (see OIL). The Greek word for "fat" is "lipos" and to the modern chemist, oils and fats are lumped together as *lipids.* (Actually, this is a name settled on only recently and one will still find lipids spoken of as *lipoids* or *lipins.*)

The third group of substances in foods contained nitrogen atoms, which lipids and carbohydrates did not. The white of egg was the best example of a food containing this nitrogenous matter. (It contains that and water and little else.) The nitrogenous matter was therefore named in its honor and called *albumin,* from the Latin "albus" (white).

Early experiments on diet proved that albuminous substances were the most essential of the three. Dogs fed on carbohydrates and lipids alone died in about a month. The German biochemist Gerardus Johannes Mulder suggested, in 1839, that albuminous substances be therefore called *proteins,* from the Greek "proteion," meaning "in first place." To this day, though, certain comparatively simple proteins, including the one in white of egg, are called albumins. The one in egg white is *egg albumin.*

Proton

In 1815, during the very early days of the modern atomic theory, the British chemist and physician William Prout suggested that all atoms were built up out of hydrogen atoms. Carbon atoms, for instance, weighed exactly twelve times as much as hydrogen atoms and were therefore made up of twelve hydrogen atoms apiece. Oxygen atoms weighed sixteen times as much as hydrogen atoms and so on. Prout suggested that hydrogen, as the primary substance out of which all else was built, be called "protyle," from the Greek words "protos," meaning "first," and "hyle," meaning "matter."

As more information was gathered, it came to seem obvious, however, that Prout's notions were wrong. The chlorine atom, for instance, was 35½ times as heavy as the hydrogen atom and, at the time, chemists were certain there was no such thing as half a hydrogen atom.

Nevertheless, in 1896 and thereafter, it was discovered that atoms were made up of still smaller particles. Over 99.9 per cent of the mass of a hydrogen atom, it turned out, consisted of a single tiny particle located at the very center of the atom. The centers of atoms heavier than hydrogen were found to contain varying numbers of this particle, so that they were made up of hydrogen atoms (in a way) after all. An atom of chlorine turned out to be 35½ times as heavy as hydrogen, because chlorine was made up of two varieties of atoms. One kind weighed 35 times as much as hydrogen atoms; the other kind, 37 times as much. Since the first type was three times as numerous as the second, the average weight was 35½.

In 1920, the British physicist Ernest Rutherford suggested that this centrally-located subatomic particle be called a *proton*. This was a deliberate tribute to Prout's "protyle," with the substitution of an "-on" suffix since that suffix had become conventional for particles within the atom.

Protoplasm

In 1665, the English physicist Robert Hooke described the tiny holes in cork as *cells*. This is a perfectly good name for holes, since it comes from the Latin "cella," meaning "a small room," or, more generally, any small hollow place. However, when later investigators, using microscopes, found that other plant and animal tissues were made up of small units marked off from one another, they persisted in calling them cells, too, though the units in living tissues, as opposed to dead tissues such as cork, were not empty.

In 1839, the German physiologist Theodor Schwann and the botanist Matthias J. Schleiden established the "cell doctrine": that is, the thesis that all living tissue was made up of cells and that the individual cell was the unit of life.

Plant cells (but not animal cells) are surrounded by "cell walls" that contain a fibrous material that was named *cellulose*, because of its place of origin. The "-ose" ending is used for sugars and related compounds and fits here, since the cellulose molecule can be broken down by acid into simple sugar molecules (see GLUCOSE).

Inside all true cells is a little spherical body called the *nucleus*. This is a diminutive form of the Latin "nux" (nut); that is, "a little nut."

The material that fills the cell is *protoplasm*. The word was first used by the Czech physiologist Johannes E. Purkinje, in 1840, and applied to the material composing young animal embryos. It comes from the Greek "protos" (first) and "plasma" (a molded form); it was the first form into which the animal was molded, in other words. The German botanist Hugo von Mohl first applied it, in 1846, to the material within the cell (which is really the first form into which an animal is molded, since we all start as a single cell).

The word is, however, falling out of use now, since protoplasm is not a single substance but rather a complex mixture of various different things, and it is these constituents of protoplasm that interest biologists and biochemists, rather then protoplasm as a whole.

Psychology

Psyche is the subject of one of the most beautiful of the Greek myths. She is a young girl with whom Eros (the god of love) falls in love. Eros marries her without allowing her to see him. Goaded on by her jealous sisters, Psyche tries to see Eros by candlelight and he forthwith leaves her. To win him back, Psyche is forced to undergo many trials and dangers but, in the end, succeeds. She is converted into a goddess and joins Eros in endless heavenly bliss.

Like most Greek myths this is really an allegory — that is, a story in which the characters and events are symbols of something else. For instance, Psyche can represent anything that undergoes trials before winning a victory that consists partly of being changed into a new and more glorious form. The ugly duckling becomes a swan; the caterpillar becomes a butterfly. The latter case was probably in the minds of the Greeks, for Psyche is usually represented in their art as having butterfly wings. (This custom persists to this day, for the fairies in storybooks of today are usually represented with butterfly wings; angels, on the other hand, have bird wings.)

However, Psyche really symbolized the human soul which, through the days of life, wearies in hardship and labor but then, at death, breaks out like a butterfly from a cocoon into a new and heavenly existence. In fact, the Greek word for "soul" is "psyche" (without the capital).

The psyche or soul includes that part of the human being that is not flesh and blood. Modern scientists have applied the word *psyche* to a man's intellect, emotions, temperament, and personality. The study of these things is, therefore, *psychology*.

A person who studies the mental and emotional processes is, generally speaking, a psychologist, but one who studies it from the medical point of view particularly and is mainly interested in mental disease, is a *psychiatrist* (the Greek "iatros" meaning "physician").

Pterodactyl

In Walt Disney's cartoon feature *Fantasia*, one of the most dramatic sequences showed a fight between a Tyrannosaurus Rex (see dinosaur) and another giant reptile. The other was a tiny-headed plant-eater some 30 feet long with spikes at the end of his tail and a series of triangular bony outgrowths standing on end in a double row down his backbone. When the bones of this creature were first found, the large triangular bones didn't seem to fit anywhere in the skeleton and at first people thought they covered the beast flatwise like the slates on a roof. The creature was named *Stegosaurus*, from the Greek "stegos" (roof) and "sauros" (lizard). It was the "roofed lizard."

The *Triceratops* was a reptile that resembled a giant rhinoceros. Its huge skull was extended into a frill covering the nape of his neck, while two horns grew over his eyes and a third of his nose. The name comes from the Greek "trikeratos" (three-horned).

The most curious of the ancient reptiles were, perhaps, those that had developed the ability to fly. Some of them were the largest flying creatures ever to live. They had long, rather narrow, leathery membranes stretched out to the end of an enormously enlarged fourth finger, the remaining fingers being small and free. (Bats, in contrast, have their flying membranes stretched across four fingers, with only the thumb free.) Such an animal is a *pterosaur*, from the Greek "pteron" (wing). It was a "winged lizard."

A small variety was called a *pterodactyl*. The Greek "daktylos" means "finger," so it was a "winged finger," which is just about right. A larger variety is the *pteranodon*, which had a wingspread of as much as 15 feet. Its skull was stretched backward into a long, narrow keel which served perhaps to keep it steady in flight, and it had no teeth. Since the Greek word for "without teeth" is "anodon," the creature's name is the curtly descriptive "wings, no teeth," though that would apply to modern birds as well.

Pyrite

BEFORE THE days of matches, one of the ways of starting a fire was to strike iron against rock. The resulting friction heated the metal to glowing heat (helped along by chemical combination of iron and oxygen), while the force of the rock broke off small pieces of the heated metal, sending sparks flying. If the sparks could be made to land in a nest of dry, tindery wood, then, with luck, the fire would be started. (We still use this principle in cigarette lighters.)

Naturally, not every rock would do in this early fire-making device. It is not surprising that a rock that turned out to be appropriate would be named in accordance. Such a rock is *iron pyrites* or, simply, *pyrite*. In Greek "pyrites" means "of fire," from "pyr," meaning "fire." Pyrite was a compound of iron and sulfur that had a yellowish metallic cast to it. The name is now applied to any sulfur-containing ore with a metallic cast.

There are copper pyrites, for instance, and tin pyrites. The former is called *chalcopyrites,* from the Greek "chalkos" (copper), and the latter *stannite* from the Latin "stannum" (tin).

But iron pyrites remains the most famous for a rather dramatic reason. Its yellowish metallic cast invariably deluded a certain percentage of the inexperienced prospectors who rushed to the California and Alaskan gold fields. Gold is hard to find, but iron pyrites with its golden glint is easy to find and many a luckless amateur went rushing in with a sack of the pyrite to stake a claim. He received scant sympathy from the old-timers whose point of view is shown in the popular name given to iron pyrites: *fool's gold.*

Minerals can fool even experts, though, and some have received their scientific names with that in mind. There is a rock called *apatite*, from the Greek "apate" (deceit) because it is so easily mistaken for other minerals. (Bones and teeth, by the way, are made up of a kind of mineral related to apatite. Oddly enough, false teeth, which are intended to deceive, are not.)

Quantum

In the nineteenth century, physicists were quite interested in the manner in which a heated body radiates energy, if only because the stars (which are heated bodies) radiate energy and it is that energy which is our only source of knowledge of the outer universe. To simplify matters, they imagined a so-called *black body*, an object which would absorb all the radiation falling upon it and reflect none, therefore appearing black. If such a body were heated, there was reason to think that some relatively simple rule ought to govern the manner in which it emitted energy.

Actual experiments with objects that were nearly black bodies showed that energy was radiated mostly in a certain narrow range of frequencies. (Radiant energy is in wave form, and *frequency* refers to the number of wave vibrations per second.) Less energy is radiated at frequencies both higher and lower. If the temperature of the body is raised, more energy of every frequency is emitted, but the point of maximum emission shifts to higher frequency.

Unfortunately, despite expectations, no reasonable theory seemed to account for the exact manner in which energy was radiated, and physicists were nonplused. Finally, in 1900, the German physicist Max Planck advanced a completely new idea. He supposed that energy, like matter, was made up of tiny particles. He showed that such energy particles would vary in size according to the frequency of a particular energy wave. The size of the energy particle, divided by its frequency, always gave the same figure, called *Planck's constant* ("constant" because it never varied).

The energy particle was called a *quantum* (plural, *quanta*), from the Latin word "quantus," meaning "how much?" and Planck's theory is called the *quantum theory*. Because quanta do behave like particles, they are also called *photons*, from the Greek word "phos," meaning "light" (which gives rise to the common prefix "photo" in scientific words), and the usual "-on" suffix used for subatomic particles.

Radar

Under the proper conditions, sound waves will be reflected from a hillside or other such obstruction. Sound travels only at the rate of about one fifth of a mile per second in air, so it usually takes a perceptible interval of time for sound to travel to the hillside and back. If the obstruction is 1100 feet away, there is a two-second interval between the sound and its reflection (which is called the *echo*, from the Greek "echos," meaning "sound"). In fact, by timing the interval between sound and echo, you can estimate the distance to the hillside.

You can do the same with any other form of radiation that can be emitted, reflected, and detected. For instance, you could send out a beam of light to a distant mirror, detect it on its return and determine the mirror's distance that way. (Of course, light travels 186,272 miles a second, so to go 93 miles and back would take only a thousandth of a second.)

During World War II, the British worked out a practical application of this long-known principle. They used short radio waves instead of light, since radio could penetrate fog, clouds, and other obstructions that ordinary light could not. The radio waves traveled at the speed of light but there are electronic devices that will determine the interval between emission and return.

Using this radio-echo principle, the British were able to detect German planes on their way to bomb London, long before the enemy were near their target. The outnumbered RAF always seemed to the puzzled Germans to be lying in wait at the right time and never to be surprised. It was radio echoes more than anything else that won the Battle of Britain.

Since the radio waves were used to tell the direction in which to send the RAF planes and the distance to send them (their range of flight, in other words), the device was called "*ra*dio *d*irecting *a*nd *r*anging," and from the initials, as italicized, the word *radar* was coined.

Radical

MATHEMATICIANS speak of roots of a number. The number 2, for instance, is the square root of 4 and the cube root of 8 (see SQUARE ROOT). The Latin word for "root" is "radix" (preserved by us today in the word *radish* which is an edible root). Any mathematical expression which contains roots is said to contain a *radical*. The sign of the radical is $\sqrt{}$.

To find the root of a number by mathematical computation is spoken of as extracting a root. The word *extract* comes from the Latin "ex-" (out) and "trahere" (to draw) and that, after all, is what is done to roots. To be made use of, they must first be "drawn out" of the ground.

The root is the basic part of the plant, or of anything — the part from which plants draw their sustenance, or ideas their significance. Any person, therefore, who wants to change the basic principles of something is a *radical*.

The word is used, then, in both mathematics and in political science, with meanings that have apparently no connection until the derivation is considered. In chemistry, the word exists in still a third completely different meaning.

In the late 1700's, the French chemist Guyton de Morveau talked about that portion of an acid other than the oxygen (in those days it was wrongly thought that oxygen was the element characteristic of acids; see OXYGEN) as the radical. It was the "root" of the acid, in other words, out of which the acid itself was built by the addition of oxygen.

The notion didn't last, but meanwhile it meant that the word *radical* had been applied to a group of atoms in a molecule. In the early 1800's the French chemist Joseph Louis Gay-Lussac took to using the term to mean any group of atoms that could be passed from molecule to molecule, during chemical reactions, as a unit. That is the meaning of *radical* in chemistry to this day.

Radioactivity

In Latin a spoke of a wheel is "radius." The same word in both Latin and English also applies to any line connecting the center of a circle with a point on its circumference. Radii (the plural form) can be drawn from the center to all points on the circumference, emerging from the central point in every direction. Such a family of lines are said to *radiate* from a point.

Light behaves in this fashion. It emerges from a candle flame or an electric bulb, traveling in straight lines in all directions. Light, therefore, is an example of a *radiation.* There are other kinds of radiations, some of which can't be detected by our senses.

For instance, in 1894, one set of radiations, imperceptible to us and much less energetic than light, was made use of by the Italian inventor Guglielmo Marconi to transmit code messages. At the time, the usual instrument for transmitting messages was the telegraph, which used wires. Marconi's instrument used radiations instead, so it was eventually referred to as a wireless telegraph or a radiotelegraph. These names were too long. The British shortened one to *wireless* and we the other to *radio.*

In 1896, uranium was found to give off new types of radiation much more energetic than light. The French scientist Pierre Curie and his Polish-born wife, Marie Sklodowska Curie, suggested, in 1898, that this phenomenon be named *radioactivity.*

The Curies discovered a new element that same year which gave off radiations much more strongly than uranium did. They, therefore, named the new element *radium.* In 1900, the German physicist Friedrich Ernest Dorn, discovered a gas released by radium during its breakdown. That gas is now called *radon.*

The Greek word for "ray" is "aktis" (genitive, "aktinos") and that was used, too. When the French chemist André Debierne discovered a new radioactive element in 1899, he called it *actinium.* Then, when the German physicists Otto Hahn and Lise Meitner discovered still another, which broke down and changed into actinium, they called it *protactinium* (that is, "first actinium," from the Greek word "protos," meaning "first").

Rayon

Nowadays, with artificial fibers numerous and varied, we tend to think of them as something new, but the first such thing was prepared over a century ago. Chemists' first attempts to imitate the fibers of nature began with the natural cellulose fiber, cotton. If cotton is treated with a mixture of *nitric acid* and *sulfuric acid* (so called because they are acids that contain a nitrogen and a sulfur atom in their molecules, respectively), nitrogen-containing groups attach to the glucose units of the cellulose, as many as three per unit. The resulting *nitrocellulose* still looks like cotton but is explosive and is called *guncotton* therefore.

If cellulose is treated less thoroughly (only two nitrogen-containing groups being attached per glucose unit), what results is not explosive, though still dangerously inflammable. It is called *pyroxylin* from the Greek "pyr" (fire) and "xylon" (wood). It is a variety of wood (i.e., cellulose), in other words, that is easily set on fire.

Pyroxylin, unlike cellulose, will dissolve in certain liquids — in a mixture of alcohol and ether, for instance — to form a thick, sticky solution called *collodion*, from the Greek "kollodes" (gluelike). If collodion solution is forced out of small holes, jets of liquid spray out. The alcohol and ether evaporate almost at once, leaving a fine fiber of pyroxylin. This first artificial fiber was patented in 1855.

Since then other less inflammable varieties of chemically treated cellulose have been used. The most important fiber resulting has been *rayon.* The name is a made-up one, implying a "ray of light" since the fibers are glossy and shiny. Such cellulose can also be forced through narrow slits to make transparent, flexible films which may be used in photography (the word film has become a slang term for motion pictures) or, if thin enough, *cellophane,* from the Greek "phanein" (to appear). In other words, cellophane is made from a substance that has the "appearance of cellulose" but is not. Rayon and cellophane both date back to the 1890's.

Relativity

IF YOU'RE sitting down now, you're not really at rest. You are whirling about the earth's axis, and the earth itself is moving about the sun, and the sun is moving about the center of the galaxy, and the galaxy itself is moving in some direction. But in the nineteenth century, physicists considered space to be filled with something called ether which they thought did not move. So they tried to determine the earth's motion through the ether to get an idea as to what its "real motion" was.

In 1887, the American physicist A. A. Michelson tried to do this by a most ingenious experiment that should certainly have worked if there were an "ether." It did not work!

In 1905, a 26-year-old Swiss patent clerk, Albert Einstein, worked out a theory to account for this. There was no ether, he said; there was nothing in the universe that was motionless. There was no such thing as "rest" or "motion" all by itself. Rest or motion depended on comparing one object with another.

The motion of the moon took it about in an ellipse if it were measured relative to the earth, something else if it were measured relative to the sun, and it was motionless if its motion were measured relative to itself. (The word *relative* comes from the Latin "re-" (back) and "latus" (carried). If motion is relative, you can only measure it by "carrying back" your observations to another body and making a comparison of motions. Human relatives are such because their blood line can be "carried back" to a common ancestor.)

But does the moon "really" move and if so, how? Einstein insisted you could not tell. From this insistence that only relative motion existed, plus the further assumption that the speed of light in vacuum was always the same, Einstein worked out a new system of the universe that explained many things older systems did not. Because relative motion was what Einstein began with, his system is referred to as *the theory of relativity*.

Rhinoceros

THE RHINOCEROS has a particularly thick hide and is really the best example of the pachyderm (see PACHYDERM), though that term is now restricted almost exclusively to elephants. There have even been stories that the rhinoceros hide is bulletproof, which it isn't. This may have originated in the fact that some species have loose skin that settles into large folds and makes the rhinoceros look as though it were encased in armor plate.

Actually, a more unusual feature of the mammal is its possession of one or two large horns on its snout. (This horn is not like the horns of other animals but develops out of glued-together hairs, so that it is more like a misplaced hoof.) The very name *rhinoceros*, in fact, comes from the Greek "rhinokeros" — "rhis" (nose) and "keras" (horn). It is the "nose-horned" creature.

There is a theory that ancient explorers brought back tales of this strange one-horned animal to Europe and that from this, plus a healthy imagination, the Europeans evolved stories of a beautiful horselike creature with one tightly spiraled horn growing straight out of its forehead. This is the mythological *unicorn*, from the Latin "unus" (one) and "cornu" (horn) — the "one-horn." It was pictured mostly in heraldry and still exists on the coat-of-arms of the British royal family.

Samples of the horn itself were passed around in more credulous times and endowed with all sorts of magical antipoison properties. Sometimes such horn proved to be rhinoceros horn, sometimes parts of the tusk of the narwhal, which is as straight and as tightly spiraled as the horn pictured on the unicorns of heraldry.

The *narwhal* is a kind of whale, the "nar-" prefix coming from a Scandinavian word meaning "corpse," in allusion, perhaps, to the sickly pale color of the creature's skin. The narwhal seems to have a unicorn horn growing out of its head, but this "horn" is really the creature's single tooth (usually from the left side of its mouth) which can be as much as eight feet long. The narwhal's scientific name is *Monodon monoceros*, from the Greek "monos" (single), "odous" (tooth), and "keras" (horn); hence, it is the "One-tooth one-horn," a pretty good name, at that.

Rh Negative

THERE IS a common Indian monkey, given the name of *rhesus* by the French naturalist Jean Baptiste Audebert in 1797. Audebert insisted that the name was simply made up and meant nothing and yet. . . .

In 1900, the Austrian physiologist Karl Landsteiner discovered that human blood might contain one of two substances, or neither (or, as was discovered two years later, both). The substances were called simply A and B so that four blood types — A, B, O, and AB — were possible. Blood also contained antibodies (see ANTIBODY) for the substance or substances it did not possess, so that B blood, for instance, could not be added to A blood or vice versa without causing the blood corpuscles to stick together and grow useless. It was only after Landsteiner's discovery, therefore, that blood *transfusion* (from the Latin "trans," meaning "over," and "fundere," meaning "to pour") became practical; and physicians knew enough to "pour over" blood from a well person to a patient that needed blood without killing the patient.

Blood substances not interfering with transfusion also exist. One was discovered in 1940 by Landsteiner and the American physician Alexander S. Wiener in the blood of a rhesus monkey. The new substance was therefore called *Rh* from the first letters. Some eight varieties of this factor are known today. No natural antibodies can exist against Rh, but in the case of all but one variety, antibodies can be developed artificially. The exceptional variety is called *Rh negative*, the others, *Rh positive*.

It sometimes happens that a mother with Rh negative blood is carrying an unborn child who has inherited Rh positive from the father. Some of the child's Rh positive may filter across to the mother's blood, which may then develop antibodies against it. If these antibodies filter back into the child's blood, they may ruin enough red corpuscles to allow a very sick baby to be born. Physicians then have to replace the baby's blood with fresh blood quickly and, to be prepared for that possibility, expectant mothers are routinely typed for Rh these days, so that at least one third of the made-up name *Rhesus* has become very significant indeed.

Science

THE GREEKS were the first who searched for understanding of themselves and the world apart from religion. They called this search "philosophia" (in English, *philosophy*), from the Greek "philos" (loved) and "sophia" (wisdom). They searched, because they "loved wisdom."

The most famous philosophers of all time, Socrates and Plato, concerned themselves with questions such as "What is virtue?" and "What is justice?" This is *moral philosophy*. The Latin "mores" means "manners" so that moral philosophy deals with man's manner of living.

But the search for understanding leads everywhere and may well concern itself with any object in nature, from the universe to a blade of grass. *Nature*, by the way, comes from the Latin "nasci," past participle, "natus" (to be born). It includes, therefore, all that "has been born," all that "has come into being" — all creation, in short.

The type of philosophy that deals with nature, rather than with the inner soul, is *natural philosophy*. To Plato, this seemed an inferior or second-class branch of philosophy, and such was his intellectual influence that for many centuries that attitude stuck.

Now, one way of avoiding a stigma is to choose a new name, or, if one must placate the gods, to use an auspicious word for an inauspicious one — a process known as *euphemism*, from the Greek "eu-" (well) and "phanai" (to speak); a word, in other words, that "speaks well" of its subject. The euphemism that came into use for natural philosophy was *natural science* or just *science*, from the Latin "scientia" (knowledge).

This, by the way, explains the very famous quotation from Hamlet: "There are more things in heaven and earth, Horatio/Than are dreamt of in your philosophy." Now Horatio was not a philosopher in the modern sense. He was a very rational-minded student of "natural philosophy" at the University of Wittenberg and he didn't believe in ghosts even when he saw one with his own eyes. Change "philosophy" in the quotation to "science" and it will make more sense.

Even today, scholars who qualify for the doctor's degree in one of our modern sciences, get the degree of Doctor of Philosophy.

Skeleton

To CHILDREN (and some adults) a set of human bones is something so frightening that it is used as one of the symbols of Halloween. It is such a caricature of a human being, with its arms and legs that are too thin and its fingers and toes that are too long; with its slatted chest, its grin, its hollow eyes. It is like a human being that has completely shriveled and dried up. The Greek word "skeletos" means "dried up" and hence, *skeleton.*

The word *skull* despite its similarity in sound, has no relationship to *skeleton.* It comes from the same Anglo-Saxon word from which *shell* is derived. Just as a soft egg or nut is surrounded by a hard shell, so is the soft brain surrounded by a hard skull. The Greek word for skull is "kranion" from which we get our word *cranium.*

In general, various bones have both common Anglo-Saxon names, like *skull,* and fancier classical names, like *cranium.*

The classical name may be derived from the bone's appearance. The collarbone, for instance, is the *clavicle,* from the Latin "clavicula" (little key), because it is a long bone which bends at the end, so that it looks like an old-fashioned key. The shoulder blade is a flat bone which is called the *scapula,* a Latin word which comes from the Greek "skaptein" (to dig) because it resembles the business end of a spade.

The name of a bone may also come from the part of the body in which it is found. The thighbone is the *femur,* which is the Latin word for "thigh" and the breastbone is the *sternum,* from the Greek "sternon" (chest).

Again, the large curving bone that makes up the main portion of the hipbone is the *ilium,* which is the Latin word for "flank." It is connected to the sacral vertebrae (see VERTEBRA) and the combined bones are referred to commonly as the *sacroiliac.* This is the Latin name most familiar to the general public, as far as the bones are concerned, because (alas) of the back pains that sometimes occur in that region.

Solstice

THE EARTH's axis is not perpendicular to its plane of rotation, but is tilted about 23½ degrees, and retains that orientation as the earth revolves about the sun. As a result the axial tilt moves first toward the sun, then away from it, and repeats the process each year.

To people on earth, it seems as though the noonday sun creeps higher in the sky each day (as one can tell by the shrinkage of the shadow of some vertical structure) until it reaches a maximum height. After that, from day to day, the noonday sun sinks lower and lower (and the shadow lengthens) until it reaches a minimum height. Then it repeats.

Primitive man naturally watched the sun's sinking with apprehension. (As it sank, winter came on and life grew hard.) Naturally, he had no assurance that the sun would ever start going up again. When the day arrived that the sun stopped sinking and turned to rise, there was general rejoicing and holiday.

The day of the year on which this happens is December 21, and this is the *winter solstice.* The word *solstice* comes from the Latin "sol" (sun) and "sistere" (to stand still). It is the day the "sun stands still." June 21 is the *summer solstice,* when the sun, having reached its highest point, "stands still" and begins to sink again.

On December 21, the sun is at zenith 23½ degrees south of the equator. That latitude is the *Tropic of Capricorn* because the sun is then in the constellation Capricorn (see ZODIAC). The word *tropic* comes from the Greek "tropikos" which in turn comes from "trepein" (to turn). It is at that point, you see, where the sun "turns" and reverses its movement.

On June 21, the sun is at zenith 23½ degrees north of the equator. This is the *Tropic of Cancer,* the Sun being then in the constellation of Cancer. The region of the earth's surface that lies between Cancer and Capricorn is called the *tropic zone,* or just the *tropics.*

Spectrum

IF A BEAM of light passes from air into glass at an acute angle, it is bent or *refracted,* from the Latin "re" (back) and "frangere" (break); the beam is "broken (or bent) back." If the glass is in the form of a triangular prism, the light on emerging is refracted farther in the same direction.

Sunlight consists, actually, of a mixture of light of varying wave lengths. These affect our eyes differently, so that we see the components of the mixture as colors. The different colors are refracted by different amounts. Red light is refracted least; orange, yellow, green, and blue are refracted in increasing amounts; violet is refracted the most.

The result (as was first noted by the English physicist Sir Isaac Newton, in 1672) is that a beam of light that has passed through a prism and been allowed to fall on a white surface becomes a "rainbow" of varying colors: red at one end, through orange, yellow, green, and blue to violet at the other.

Since the colored strip is pure light, it is called a *spectrum* which is the Latin word for "image" or "apparition."

Particular substances, when heated to a white heat, give off only certain colors. If the radiating light is passed through a slit, each color will form a sharp image of the slit, falling in a certain position in the spectrum, leaving the rest black. On the other hand, sunlight passing through a cool gas will have certain of its colors absorbed and dark lines will appear against a colored background. The outer layers of the sun are cool enough to do this so that the solar spectrum is actually crossed by dark *Fraunhofer lines*, so called after the German optician Joseph von Fraunhofer, who first observed them in 1814.

An instrument through which one can view a spectrum against a marked scale so that the position of each bright or dark line can be located exactly is a *spectroscope.* From the position of the lines, mankind can and has learned the composition of the sun and the stars.

Spiral

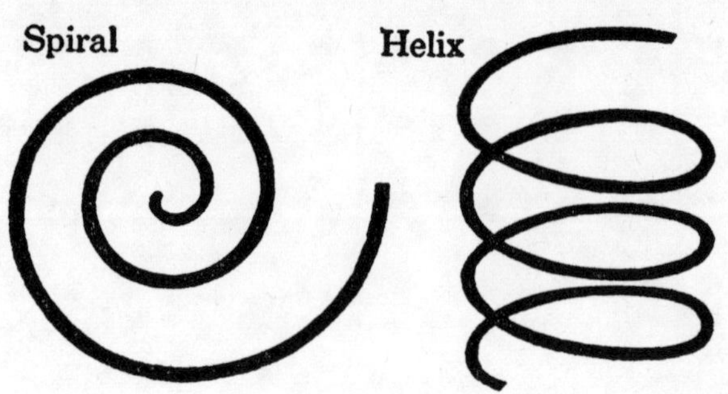

Curves that lie in a plane and are two-dimensional have rather familiar names: circle, ellipse, oval, and so on. Partly, this is because such curves are easy to show on a blackboard or on the page of a book and are thus more frequently talked about. Curves which extend through all three dimensions are a little less familiar.

Imagine trying to draw a circle in the air with your finger and simultaneously moving your finger away from you. The path made by your finger is like that of the groove on a screw. Such a screw-like curve is called a *helix* in both Greek and English.

The helix is important in physics because wires wrapped around an iron core to form an electromagnet follow such a path. A magnet can be moved through a helix of wires to set currents going. The emphasis in such a wire helix is usually on the channel within the coil through which the magnet can be moved. Such a helix is often called, for that reason, a *solenoid*, from the Greek "solen," meaning "channel."

The Greeks had another word for "coil" or "twist" and that was "speira" which may have been derived from "sparton," meaning "rope." The common rope, after all, is formed by coiling or twisting a number of fibers about a common axis. The fibers, when so coiled, automatically take up a helical shape so the word *spiral* is often used as a synonym for *helix*. For instance, a stairway that curves upward in a helix is called a spiral stairway rather than a helical stairway.

But *spiral* has an additional meaning. A common example of a helix or spiral in nature is the snail shell, which coils round and round as the snail grows. But as the snail grows, the size of the shell grows in such a way that, in coiling about, it makes larger circles with each coil. For this reason, *spiral* has come to mean a two-dimensional curve which recedes from a center as it curves (like a snail shell seen from the side).

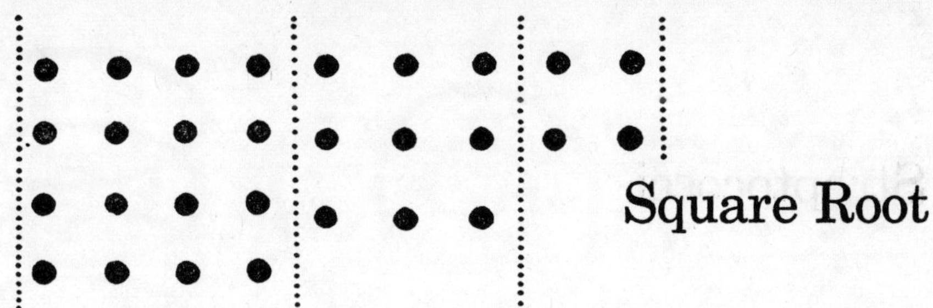

Square Root

The ancient Greeks loved to arrange dots in geometrical shapes and count up the number of dots it took to make triangles, squares, and so on. A perfect square could be made with four dots (two on each side), or with nine dots (three on each side), sixteen dots (four on each side) and so on. For this reason, numbers like 4, 9, 16, 25, 36, 49, etc. were termed *squares*. For each square, there was a "side" ("latus" in Latin). The side of 4 was 2; the side of 9 was three; the side of 16 was 4 and so on.

If you forget about sides and squares, you might concentrate on the fact that 2 × 2 is 4; 3 × 3 is 9; 4 × 4 is 16 and so forth. You can therefore define a square as any number that results from the multiplication of a digit by itself. And the "side" of a square is the number which, when multiplied by itself would give the square.

The medieval Arabs, who studied and helped to preserve Greek mathematics during the European Dark Ages, were less interested in geometrical figures than were the Greeks and more interested in arithmetical relationships. To them, if 2 × 2 was 4, then 4 was built up out of 2's as a plant is built up out of a root. They didn't think of 2 as the side of a square but as the root of something that was larger. They called it that and we call it that to this day.

Now dots can be arranged in cubes, also. Eight dots can be arranged in a cube, two on each side. Twenty-seven dots can be arranged in a cube, three on each side, and so on. The Greeks called numbers like 8, 27, 64, and 125 *cubes* — and we do, also. Arithmetically, cubes are obtained by multiplying a number by itself twice: 2 × 2 × 2 is 8; 3 × 3 × 3 is 27 and so on. So 2 is a root of 8 as well as of 4, and 3 is the root of 27 as well as of 9. To distinguish between these roots, it is only logical to speak of *square roots* and *cube roots*. The number 2 is the square root of 4 and the cube root of 8.

You can extend this even further, so that 2 is the fourth root of 16, the fifth root of 32, and so on.

Streptococci

THE TERMS *microbe* and *germ*, often applied to bacteria, are really too broad for this use (see MICROBE) and, strangely enough, the term *bacteria* is too narrow. It is derived from the Greek "bakterion," meaning "a little rod," but not all bacteria have the shape of little rods. In fact, a bacterium that does have that shape is now referred to as a *bacillus*, from the Latin meaning "a little rod," a diminutive of "baculus," meaning "staff."

The question of names has always plagued bacteriologists anyway, simply because they deal with creatures that are so small, mere blobs of life.

Nevertheless, names have been applied to certain broad groups on the basis of differences in appearance. For instance, many bacteria are not rodlike but are simply tiny spheres. Such a bacterium is a *coccus* (plural, *cocci*, pronounced "cock's eye"), from the Greek "kokkos," meaning "grain" or "seed." Sometimes they are called *micrococci;* the Greek "mikros" means "small."

Cocci sometimes form pairs with a common cell wall about the two, and these are *diplococci* (from the Greek "diploos," meaning "twofold"). One of these causes some types of pneumonia and is sometimes referred to as *pneumococci*, for that reason.

Some varieties of cocci divide in such a way that the daughter cells remain attached, each dividing in its turn so that, eventually, long strings of them are formed which, as strings will, may curve or twist. These are *streptococci*, from the Greek "streptos," meaning "twisted." One variety of these can give rise to throat infections, hence our slang expression, "strep throat."

If cocci divide in such a way that descendants stick together in a bunch rather than simply in a string, they are called *staphylococci*, from the Greek "staphyle," meaning "a bunch of grapes." Ordinary boils are staphylococcal infections.

Rodlike bacteria that twist into coils are *spirilla* (a Latin word meaning "a little coil," from "spira," a coil, see SPIRAL). The word *spirilla* was actually made up after Latin ceased to be a living language.

Sulfanilamide

AN ATOM combination made up of a sulfur atom, three oxygen atoms, and a hydrogen atom is known as a *sulfonic acid group.*

Now an amine group (see AMMONIA) which replaces an oxygen and a hydrogen atom of an acid group is given the special name of *amide.* The special name is convenient because an amine group attached to an acid group has properties that are different from those of an amine group attached elsewhere.

A sulfonic acid to which an amine group has attached itself is, therefore, called a *sulfonamide group.* If a sulfonamide group is attached to a molecule of aniline (see ANILINE), the resulting compound is given the composite name *sulfanilamide* (*sulf-anil-amide*).

Sulfanilamide was first synthesized in 1908 but for nearly 30 years afterward was important only as a component of dye compounds. One such compound was Prontosil. In 1934, the German chemist Gerhard Domagk found that Prontosil was amazingly effective in clearing up certain types of infections. (In 1939, he was awarded the Nobel Prize in medicine for this but was forced by Germany's Nazi government to decline it.) In 1935, French chemists showed that it was the sulfanilamide portion of Prontosil that did the trick.

At once, there was a mad scramble to prepare other drugs related to sulfanilamide. These new drugs, it was thought, might be even more effective than sulfanilamide, or might attack some kinds of germs that sulfanilamide did not. Over 5000 such drugs were prepared in the next ten years. For the most part changes were made by attaching various types of atom groupings to the amide group of sulfanilamide. When the chemicals pyridine, thiazole, or diazine were attached, for instance, the resulting compounds were named *sulfapyridine*, *sulfathiazole*, and *sulfadiazine*, by analogy with sulfanilamide.

This was false analogy because sulfanilamide is a name made up of "sulf-anil" and not "sulfa-nil." Nevertheless, this has persisted and the whole group of drugs is popularly known as the *sulfa drugs.*

Tangent

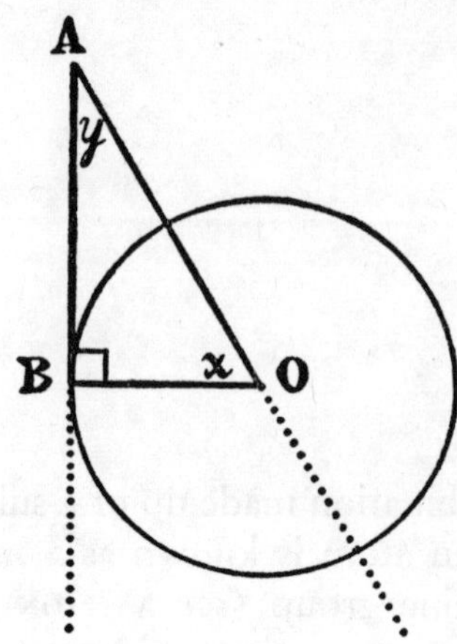

Consider the right triangle *ABO* in the accompanying figure. Line *BO* is the radius of the circle. Line *AB*, if extended, will be seen to just barely touch the circle at the single point *B*. Line *AB* is a *tangent* of the circle (from the Latin word "tangere," meaning "to touch").

The ratio of the length of line *AB* (the side opposite angle *X*) to the length of line *BO* (the side adjacent to angle *X*) is a fixed figure characteristic of this particular angle. Since this ratio is that of a tangent to the radius, it is referred to as the *tangent of angle X*, or, for short, *tan X*.

In any right triangle, angles *X* and *Y* add up to exactly a right angle, so one is said to be the *complement* of the other. This is from the Latin "complere" (to fill up), since angle *X* added to angle *Y* (or vice versa) fills up a right angle.

The tangent of angle *Y* would be the ratio of its opposite side (line *BO*) to its adjacent side (line *AB*), by analogy. Since this ratio is the tangent of the complement of angle *X*, it is referred to as the *cotangent of angle* X, or *cot* X. The prefix "co-" is made up of the first two letters of *complement.*

Referring back to the figure, you will see that the hypotenuse of the right triangle (line *AO*) is a line that cuts through the circle. It is called a *secant,* from the Latin word "secare," meaning "to cut." The ratio of line *AO* to that of line *B*O is that of secant to radius and it is called the *secant of angle* X, or *sec* X.

Arguing as before, the secant of angle *Y* (the complement of angle *X*) would be the ratio of line *AO* to line *AB* and that ratio is, therefore, the *cosecant of angle* X, or *csc* X.

Tantalum

In Greek mythology, Tantalus was a king of Lydia who had seriously offended the gods. He was therefore condemned to suffer harsh punishment in Hades. He stood in water up to his neck but when, driven by thirst, he stooped to drink, the water swirled down and out of sight. Branches laden with fruit dangled within inches of his face but when, driven by hunger, he reached out to eat, they swayed just out of his grasp. The king has left his name in our language in the word *tantalize*.

His name enters chemistry, too, as an element. In 1802, the Swedish chemist Anders Gustaf Eckeberg discovered element number 73, and thereafter, for a number of years, there was some dispute as to whether it was really a new and distinct element and, if so, what its name should be. In 1814, the Swedish chemist Jöns Jakob Berzelius, who was the great chemical authority of the day, decided it was indeed a new element and also decided in favor of the name *tantalum*.

He reasoned that the new element was unusual for a metal in resisting the action of acids, even of aqua regia (see AQUA REGIA). Though it stood in acid, in other words, it was not affected by it, any more than was Tantalus by the water in which he stood.

Some people have suggested that tantalum received its name because the discoverer had been tantalized by near misses before finding it. Though this is apparently a false derivation, there are other elements that do bear witness to their discoverers' difficulties.

For instance, the Swedish chemist Carl Gustav Mosander discovered element 57 and called it *lanthanum* from the Greek "lanthanein" (to escape notice) because he found it so difficult to isolate. This element is the first of the rare earth elements (see YTTRIUM) and, in consequence, this rather peevish name is applied to the whole series of rare earth elements, which are termed the *lanthanides*. Another of these elements (number 66) was discovered in 1886 by the French chemist Lecoq de Boisbaudran who named it *dysprosium*, from the Greek "dysprositos" (hard to get at).

Technetium

By 1925, all but four elements in the periodic table (see ISOTOPE) had been discovered. Two of these "holes" in the table were among the heavy radioactive elements and were expected to be rare and hard to find. The other two, in positions 43 and 61, were surrounded by stable elements and it seemed there ought to be no great problem in locating them.

Chemists concentrated on the search and there were a number of reports that one or the other had been detected. For instance, in 1925, three German chemists reported the detection of element 43 and named it *masurium* after Masuria, a district in East Prussia (now part of Poland). For fifteen years, periodic tables listed masurium (but with a question mark).

The next year, American chemists at the University of Illinois and Italian chemists at the University of Florence reported the detection of element 61. The first group called it *illinium,* the second *florentium,* after their respective universities. There was quite an argument about it, but American periodic tables listed illinium (again with a question mark).

It has since turned out that everyone was wrong. Elements 43 and 61 are radioactive and do not exist naturally on earth. They can, however, be formed by nuclear reactions which, since 1919, mankind has known how to bring about. In 1936, for instance, the American physicist Ernest Lawrence bombarded the element molybdenum (element 42) with subatomic particles, and investigation showed that small quantities of element 43 had been formed. Eventually, element 43 was named *technetium* (now the official name), from the Greek word "technetos," meaning "artificial," since technetium was the first element to be discovered through formation by artificial means.

In 1945, element 61 was found by American chemists at Oak Ridge among the uranium fission fragments. It was named *promethium* after the Greek demigod Prometheus, who had brought down fire to mankind from the sun. The new element, after all, had come out of the man-made sun of uranium fission.

Teleostei

ALTHOUGH *vertebrate* is a common term used to include all animals that have bony skeletons (see PHYLUM), there are actually some creatures that are quite similar to the vertebrates, similar enough to be classified with them, which yet do not have a bony skeleton. A skeleton, yes, but not of bone.

There is a time when all vertebrates have a skeleton not of bone, but of *cartilage* (from the Latin "cartilago," of uncertain derivation, which has replaced the Anglo-Saxon "gristle"). Cartilage is a tough, flexible tissue which gives support but is not particularly hard. It develops first and bone follows. A baby's skeleton is largely cartilage and hardens only slowly over the years as it turns to bone, or *ossifies* (from the Latin "os" for "bone" and "ficare," meaning "to make"). Adults retain cartilage in many places, notably in the ear and in the tip of the nose.

In some animals, the skeleton remains cartilage throughout life. The best examples are found among the animals in the class *Pisces* (the Latin plural word for "fish," which is the type of animal mainly contained in the class). *Fish* itself comes from the Anglo-Saxon "fisc," which is obviously related to the Latin word.

Ordinary fish — cod, mackerel, salmon, herring — have skeletons of bone and are sometimes called *bony fish* to emphasize that fact. Zoologists put them all in the subclass *Teleostei,* from the Greek "teleos" (complete) and "osteon" (bone). The skeletons of these fish are "completely bone."

However, included in Pisces is another subclass consisting of creatures which in some cases look very much like bony fish (with some minor but characteristic differences) but which have skeletons made up entirely of cartilage throughout life. These include the sharks and certain less fishlike relatives. These are the *Elasmobranchii,* from the Greek "elasma" (a metal plate) and "branchia" (gills) because their gills have a flat platelike structure.

Telescope

A COMMON plaything today is the glass lens which will magnify print and focus sun's rays sufficiently to scorch paper. (*Lens* is a Latin word meaning "lentil." The most common form of lens is shaped like a lentil seed.)

But lenses are much more than toys. In 1608, a Dutch spectacle-maker named Jan Lippershey placed two lenses in a tube and made distant things appear close. He applied for a patent, but the Dutch government scented "secret weapon," refused the patent, but bought all rights and told Lippershey to keep experimenting.

Secrecy, as usual, did no good. Rumor of the discovery spread and in 1609, the Italian physicist Galileo Galilei reinvented a similar instrument and began exploring the heavens. The instrument was called a *telescope*, from the Greek "tele" (distant) and "skopein" (to watch). It gave man the means to "watch the distance."

In rapid succession, Galileo discovered mountains on the moon, spots on the sun, the phases of Venus, and the four largest moons of Jupiter. The last are still called the *Galilean satellites* in his honor.

The use of "-scope" as a suffix in the names of scientific instruments that help man to see the unseen is widespread (see SPECTRUM, MICROBE, COSMIC RAY). However, man can "watch" his environment by senses other than sight, and in the case of one very familiar instrument, it is the sense of hearing that counts.

For centuries, doctors have tried to gain knowledge of what is going on within the chest by placing the ear against the chest wall to catch the noise of the heartbeat and of the pumping air. This is known as *auscultation*, from the Latin "auscultare" (to listen). In 1819, the French physician Rene T. H. Laennec devised a tube, one end of which could be placed against the chest to sharpen and make clearer the noises within to an ear placed at the other end. That was the birth of today's physicians' indispensable *stethoscope*, named from the Greek "stethos" (breast).

Terpene

THERE IS A small tree growing on the shores of the Mediterranean which the Greeks called "terebinthinos." If the bark of the tree is cut, a yellow sticky fluid wells out, which hardens after a while on exposure to air. This substance is called *turpentine,* which is merely a corruption of "terebinthinos."

Terebinth turpentine is now called Chian turpentine because it was originally collected on the island of Chios in the Aegean Sea. A more important modern source are a number of cone-bearing trees (or *conifers,* the Latin "ferre" meaning "to bear") such as pines and firs.

If crude turpentine is heated with boiling water, some of it is driven off with the steam. If this portion is trapped and cooled, an oily liquid called *spirits of turpentine* (for the meaning of *spirit,* see ETHER) results. What is left behind is a yellow-brown brittle solid called *rosin.* This word is a corruption of *resin,* which is the name given to the general class of gummy fluids that flow from the bark of trees and harden in air. The Greek word for it was "rhetine," a close relative of "rheein," meaning "to flow."

Spirits of turpentine contains a number of organic molecules, each of which contains ten carbon atoms arranged in such a way that they can be divided into five-carbon halves, with each half consisting of four carbon atoms in a line and a fifth carbon attached to the second. Such ten-carbon compounds are *terpenes,* from "turpentine."

The five-carbon unit out of which terpenes are built was called *isoprene* in 1860 by a chemist named C. G. Williams. The "iso-" prefix is often used in naming carbon chains with one carbon atom branching off a second carbon in line, but the "-prene" part seems to have been improvised and to have no special meaning. Because isoprene is half a terpene molecule, it is sometimes called *hemiterpene,* the Greek prefix "hemi-" meaning "half."

Thermonuclear Reaction

ALTHOUGH THE first practical use of energy derived from nuclear reaction involved the splitting of large nuclei (see FISSION), scientists had speculated about the possibility of energy from the joining of small nuclei long before anyone had even conceived of splitting large ones. To combine four hydrogen nuclei, for example, into one helium nucleus would liberate tremendous energy. This process is called nuclear *fusion,* from the Latin "fundere" (past participle, "fusus") meaning "to melt." Four lead pellets, for instance, if melted, would run together and form one larger pellet, if cooled again. So *fuse* came to mean the joining of many small pieces into one large one.

The trouble in achieving nuclear fusion was that the hydrogen electrons got in the way. The hydrogen nuclei just couldn't be forced close enough to fuse by anything scientists could do until the A-bomb was invented (see NUCLEAR REACTOR). The A-bomb developed a heat of about a hundred million degrees and under certain circumstances this was enough to strip away the electrons and bang the hydrogen nuclei together hard enough to cause them to fuse. An explosion much more powerful than an ordinary atomic explosion resulted.

This new type of bomb was as much an "atomic bomb" as the old A-bomb, or as little, since both are really nuclear bombs (see NUCLEAR REACTOR), but to distinguish between them, the new bomb, which fused hydrogen nuclei, was called a *hydrogen bomb,* or an *H-bomb.* There was some attempt to call the older bomb a *uranium bomb* or a *U-bomb,* but that didn't catch on. A more logical division, which is catching on, is to call the H-bomb a *fusion bomb* and the ordinary A-bomb a *fission bomb.*

Ordinary nuclear reactions, including fission, are brought about in the laboratory by bombarding atomic nuclei with subatomic particles. Hydrogen fusion, however, is brought about by sheer heat. The Greek word for "heat" is "therme," so fusion is a *thermonuclear reaction* and the H-bomb is sometimes called a *thermonuclear device.*

Thiophene

THE COMPOUND benzene is usually obtained from petroleum or from coal tar. When it is so obtained, unless special precautions are taken, small quantities of another substance accompany it. The benzene molecule, you see, is made up of six carbon atoms in a ring. The molecule of the other substance is made up of four carbon atoms and a sulfur atom in a ring. The sulfur atom apparently takes up just the space of two carbon atoms, so that the molecules are sufficiently alike in shape to respond to similar treatment. Whatever course of events isolates one, isolates the other.

Chemists were so unaware of the presence of the impurity that they used to test for the presence of benzene in a liquid by adding a bit of concentrated sulfuric acid and a crystal of a substance called isatin. The result was a beautiful blue color which did not appear unless "benzene" was present in the liquid. What the chemists didn't know for quite a while was that the impurity gave the color and not the benzene.

The German chemist Victor Meyer used to demonstrate this test to his class until one day in 1883 Meyer's laboratory assistant supplied him with some benzene prepared in a new way — from chemically pure benzoic acid. This new sample had only benzene in it and no impurity. Naturally, it did not give the test. One can imagine the scholarly Meyer staring dumfounded at the test tube and shaking it helplessly while his class longed to laugh and dared not.

Meyer did not let this drop. He investigated the peculiar happening and did not rest until he had located the impurity and worked out its structure. (The moral to this story is that the good research scientist pays strict attention to any peculiar event and does not shrug off a mystery as just "one of those things.")

The benzene twin was named *thiophene*. The "thio-" prefix, in chemical names, invariably implies the presence of a sulfur atom in the molecule, from the Greek "theion" (sulfur), while "phene" is from "pheno," one of the early names of benzene (see BENZENE). *Thiophene* therefore means "sulfur-benzene," a good and descriptive name.

Thyroid

When the Greeks of the heroic age (as in the *Iliad*) went into battle, they took care to take along a large shield behind which they could remain whenever possible. Ajax (the Grecian fighter second only to Achilles in valor) was noted for his curved, oblong shield that reached from neck to ankles. Now the Greek "thyra" means "door," so such a large shield was called a "thyreos" (carrying such a shield was like carrying a door).

Such a really large shield had a notch on top over which the hero's head could take a quick look, when necessary. If you put your fingers on your *larynx* (a Greek word meaning "throat" or "gullet") or *Adam's apple* (so called because medieval legend had it that Adam choked on the apple and part of it stuck in his throat) you will notice that it has a notch on top. It is therefore "shield-shaped" and was named the *thyroid cartilage* (the suffix "-oid" comes from the Greek "oeides" meaning "form"). Or, of course, it may simply be that the ancients thought of the Adam's apple as a door in the throat.

In any case, the name spread. A gland in the close neighborhood of this cartilage was named the *thyroid gland*. In one sense, the thyroid gland is indeed a shield, since it secretes a hormone (the most familiar form of which is called *thyroxine*) which regulates the rate at which the body burns its fuel and uses up oxygen. The gland thus shields us against improper "rate of living." The discoverer (in 1915), Dr. Edward C. Kendall of the Mayo Clinic, added the "ox" of "oxygen" and the "-in" suffix common for hormones to the "thyr" of "thyroid" and got the name.

An unfortunate child born without the ability to produce thyroxine turns into a dwarfed idiot called a *cretin*. This is a corruption of the French *chretien* meaning "Christian." It is fairly common, you see, among people who do not understand mental ailments to consider a mentally deficient person to be under the special care of the gods. Our word *silly*, for instance, is a corruption of the Anglo-Saxon "saelig" (and the modern German *selig*) which means "blessed." (Perhaps this is because mental deficients seem childishly happy at times and may seem unaware of the cares and troubles of the world. There may be blessing in that, after all.)

Tide

Just as the earth's gravity holds the moon in its orbit, so the moon's gravity has its effect on the earth. Our satellite pulls the water of the ocean toward itself a bit so that the water piles up into a kind of hill in the direction of the moon. It pulls the earth itself away from the water on the side of the earth opposite to itself so that another hill forms on that side.

If the earth were entirely covered with water, these hills would seem to move around the earth as it rotated, keeping themselves always lined up with the moon, but lagging a little because of the earth's movement. If an island existed, then, twice a day (at progressively later times since the moon was moving, too, and not staying in one place), one of the water-hills would reach it. The water level would rise and then recede again as the hill passed.

This actually does happen along shores but because the land masses get in the way and break up what would otherwise be an orderly progression of the water-hills, the rise and fall of the water doesn't follow the moon quite as closely and as obviously as it would otherwise. Also, depending on the shape of the shore, the rise and fall can make a difference of 60 feet in a funnel-shaped bay like the Bay of Fundy or only a few inches in a nearly land-locked sea like the Mediterranean. So the connection of the water-hills and the moon went unnoticed until early modern times.

What was strongly noticed, however, was the regularity of the rise and fall. Those who sailed vessels had to know when the water was high and when low and they found they could count on it in advance. So the rise and fall was named *tide*, from the Anglo-Saxon "tid," meaning "time." The tides were so regular they could be used as a measure of time. The old meaning of the word *tide* is still preserved in archaic words such as *eventide* and *Yuletide*. When we speak of "time and tide," we are using synonyms as in "spic and span," "hale and hearty," and "nook and cranny."

Trigonometry

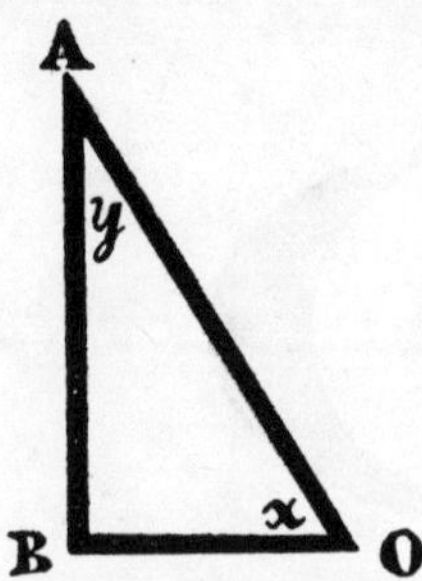

THE ANGLES of any triangle add up to 180 degrees; thus, if two angles are known, the third can be determined at once. In a right triangle, one of the angles is 90 degrees, so that if one of the two remaining angles is known, the other is also known.

Furthermore, in any right triangle with known angles, if the length of one of the sides is also known, the lengths of the other two can be calculated. For instance, in the right triangle in the accompanying figure, if angle *X* is known, then the ratio of side *AB* to side *BO* is the tangent of angle *X* (see TANGENT) and this ratio can be looked up in certain tables used for the purpose. Using the ratio, the length of *AB* can be calculated from the known length of *BO* or vice versa. And if *AB* and *BO* are both known, so that the ratio can be worked out, the size of angle *X* can be looked up in the table.

There are six possible ways of taking ratios of the sides of a right triangle. Four of these — the tangent, cotangent, secant, and cosecant — are mentioned under TANGENT. A fifth is the ratio of line *AB* (the side opposite to angle *X*) to line *AO* (the hypotenuse). This is the *sine of angle* X or *sin* X, for short. This comes from the Latin "sinus," meaning "bend" or "curve" (from which we get two other words, *sinus*, and *sinuous*), perhaps because if the value of the sine is graphed for angles of increasing size, the result is a peculiar wavelike curve called a *sine curve* or *sine wave*. The ratio of line *BO* to line *AO* is, correspondingly, the sine of angle *Y* and therefore (see TANGENT) the *cosine of angle* X or *cos* X.

The study of these ratios belongs to that branch of mathematics called *trigonometry*, from the Greek "tri-" (three), "gonia" (angle) and "metron" (measure), so that it means "the measurement of triangles," which is just what it is. The various ratios of the sides of the triangle are called *trigonometric functions*. *Function* is a general term in mathematics referring to the relationship of two quantities. Mathematicians do things with them and the Latin "fungor" (past participle, "functus") means "to perform, or execute or do."

Trypsin

Among the earliest enzymes discovered were those in the digestive juices which dissolved or liquefied meat. This ability was reflected in the naming of pepsin (see ENZYME) which is found in the stomach juices. Other digestive enzymes were later discovered and the name *pepsin* inspired the later namings.

In 1874, for instance, the German physiologist Willy Kühne discovered an enzyme in the juice formed by the pancreas which resembled pepsin in action. Since he obtained this enzyme by rubbing and grinding the pancreas in glycerol, he named the enzyme *trypsin*, a combination of the Greek "tribein" (to rub) and "pepsin." It was, in other words, a kind of pepsin obtained by rubbing. When a second and similar enzyme was found in the pancreatic juice, it was called *chymotrypsin*, the prefix "chymo-" coming from the Greek "chymos" (juice).

The inner lining of the small intestine contains small glands that produce intestinal juice. In 1901, the German pathologist Julius Cohnheim isolated still another enzyme, related in action to the ones already mentioned. This he called *erepsin* from the Greek "ereptesthai" (to feed upon) because its enzyme action was a kind of feeding upon substances from meat. (Still, I have a notion that the similarity of the name *erepsin* to *pepsin* also influenced him.)

All these enzymes work by splitting up the large protein molecules into smaller pieces. They are therefore called *proteases*. (The "-ase" suffix is now accepted as signifying an enzyme.) Eventually, the individual amino acids making up the protein molecules are set free. One important amino acid was first isolated in 1900 by the British biochemists Frederick G. Hopkins and S. W. Cole from protein that had been digested by trypsin. They named the amino acid *tryptophane*, the suffix "-phane" coming from the Greek "phainein" (to appear). It had appeared, so to speak, as the result of trypsin's activity.

Umbra

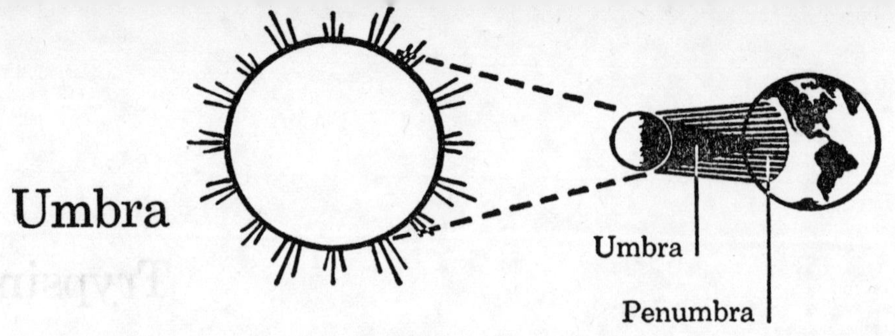

ANY DARK body in the neighborhood of a sun (as the earth or moon, for instance) casts a shadow. To an object located anywhere within the shadow, the sun is invisible.

When the moon is eclipsed, it is because it passes through the earth's shadow. When the sun is eclipsed, the earth is passing through the moon's shadow. The shadow of a world narrows as it extends outward because the world decreases in apparent size with respect to its sun as distance from it increases.

From the distance of the moon, the earth still appears to be four times the diameter of the sun, so earth's shadow is thick enough to cover the entire moon. From the same distance, however, the moon, a smaller body, seems just the size of the sun and its shadow is narrowed down to almost nothing. No sooner has it moved to cover up all the sun than, as it continues moving, the sun begins to emerge at the other side. For this reason, any particular eclipse of the sun is confined to a small region of the earth and only lasts for seven minutes at the most.

A world's shadow is its *umbra* (the Latin word for "shadow") while the region near it, where the sun is partly but not entirely covered, is the *penumbra* (from the Latin "paene," meaning "almost," hence the "almost shadow"). Similarly, the central part of a sunspot, which, though glowing hot, seems dark and shadowy against the still more glowing surface of the rest of the sun, is also called an umbra, and is also surrounded by a penumbra.

The word enters astronomy in still another way. Alexander Pope, in his poem "The Rape of the Lock," invented a moody, sorrowing spirit, who lived in the shadow of grief, so to speak, and whom he called Umbriel in consequence as a sort of counterpart to the gay spirit Ariel (the name being chosen with "airy" in mind, probably) in *The Tempest* by Shakespeare. In 1851, when William Lassell discovered the third and fourth satellites of Uranus, he decided to give them the names of spirits since the first two had been named Oberon and Titania after the king and queen of fairies. Consequently, he gave one the name *Umbriel* and the other *Ariel.*

Ungulate

THE HORNY sheaths at the ends of a mammal's fingers and toes can take on several shapes. As almost every child learns by sad experience, a cat possesses claws. On the other hand, a horse possesses hoofs. This suggests a sensible way of making a division among mammals. After all, this is a considerable difference. Claws are offensive weapons and mark the aggressive meat-eater; while hoofs are devices for running away and mark the timid plant-eater.

The English naturalist John Ray, in 1693, did make just such a classification. The Latin word for "fingernail" is "unguis" and the Romans made two diminutives of it; a small nail could be "unguiculus" or "ungula" and Ray made use of both. A clawed mammal he called an *unguiculate* and a hoofed mammal he called an *ungulate.* The two terms still exist, embracing two of the subsections of the placental mammals (see MARSUPIAL). However, allowance is now made for mammals that don't fit well in either group.

For instance, there are mammals that have nails that are neither claws nor hoofs, but actual nails like ours. Such animals include ourselves, of course, and also the rather similar-appearing apes and monkeys. The name of these animals, however, has nothing to do with "nail" in either English or Latin. The group includes ourselves and so, with simple and natural conceit, Karl von Linnaeus, the great classifier of plants and animals, called a nailed mammal a *primate,* from the Latin "primus" (first).

And then there are mammals who have so adapted themselves to the sea that they have left only two flipper-shaped forelimbs with no visible nails of any type. These include the whales, porpoises, and dolphins. A non-nailed mammal like that is a *cetacean,* from the Greek "ketos" (whale).

Birds, by the way, also have claws. The unusual thing about birds, though, is that at least one toe on each foot points backward, so there is at least one claw on a bird's foot where, on a mammal, the heel would be. A bird's claw is therefore a *talon,* from the Latin "talus" (heel).

Universe

A GROUP OF individuals acting, in combination, toward a single goal under a single direction, behaves as though it were one person. In the Middle Ages, such a group was called a "universitas," from the Latin "unum" (one) and "vertere" (to turn); a group, in other words, had "turned into one" person. In the most general sense, this came to mean the group of everything in existence considered as a unit; that is, the *universe.*

It also came to be used in a more restricted sense. In the early Middle Ages, for instance, a school of higher learning was called a "studium," from the Latin "studere," meaning "to be zealous" or "to strive after." From this, come our words *study* and *student.* The Italian version of "studium" is *studio,* which has come into English as a place where the fine arts, particularly, are studied or practiced (a memorial to the fact that in Renaissance times, Italy was the center of the world of fine art).

A group of students at such a school would refer to themselves (with the usual good opinion of themselves that students always have) as "universitas magistrorum et scholarium" — a "group of masters and scholars." They were a "universitas," you see, because they were all pursuing the single goal of learning. And gradually the name of the group became, in shortened form, the name of the schools which, around 1300, began to be known as *universities.*

Meanwhile, within the university, groups of students following a particular specialty, as, for instance, law, would band together for mutual aid. They were a group of *colleagues* (Latin, "collegae") from the Latin "com-" (together) and "ligare" (to bind). They were bound together for a common purpose. Such a group of colleagues formed a *college* (Latin, "collegium"), a word now used to denote a particular school within a university.

The older meaning as simply "a group of colleagues" persists today in the College of Cardinals of the Roman Catholic Church, and in America's own electoral college which meets every four years to elect a president.

Uranium

In 1781, the German-born British astronomer William Herschel discovered a new planet and caused a tremendous sensation in the scientific world. It was the first discovery of a planet in recorded history. To be sure, since the invention of the telescope, over 150 years earlier, four satellites had been discovered circling Jupiter, and four circling Saturn, but these were not independent worlds. This new discovery was a true planet, circling the sun at twice the distance of the farthest anciently known planet, Saturn.

This new planet (the seventh) was named Uranus, after the Greek god of the sky, Ouranos, who, according to the Greek mythology, had been the father of Cronos (Saturn, according to the Romans, and the sixth planet) and the grandfather of Zeus (Jupiter, according to the Romans, and the fifth planet).

So much for that. Now in 1789, the German chemist Martin H. Klaproth was working with a heavy black mineral called pitchblende. In it, he found evidence of a new metal, hitherto unknown. It had been an old-fashioned habit of the medieval alchemists to refer to metals by the names of various heavenly bodies. Gold was called the Sun; silver, the Moon; copper, Venus; iron, Mars; and so on (see MERCURY). Now here was a new metal and a new planet, too. Klaproth therefore named the new metal *uranium* after the new planet.

A century and a half later there was an echo to this story. In 1940, American scientists at the University of California formed two new elements by means of nuclear reactions. Previously, uranium (with an atomic number of 92) had been the most complicated known element. The two new elements, however, were more complicated still, with atomic numbers of 93 and 94. So they were named after the two planets beyond Uranus, planets that had been discovered since Herschel's day. Element 93 was named *neptunium* after Neptune (the eighth planet and the Roman god of the sea) and element 94, *plutonium*, after Pluto (the ninth planet and the Greek god of the underworld).

Vaccination

UNTIL 150 years ago, smallpox was a common and dreaded disease. It often killed (as it did Louis XV of France) or, if it did not kill, it usually disfigured (as, for instance, George Washington), leaving the face pock-marked. (The word *pock*, from the Anglo-Saxon, is applied to any skin eruption resulting from disease or to the scar it leaves behind when it is gone.)

People who got smallpox and survived did not get smallpox a second time. They were resistant to it. Moreover, even in the 1700's, it was suspected that if a person caught cowpox (a much milder disease, caught from cows, and hence its name) and recovered, that person was thereafter resistant not only to cowpox but also to the much more dangerous smallpox. (It is perhaps for this reason that milkmaids were always so beautiful in the eighteenth-century romantic comedies. They caught cowpox early in life and were spared the disfigurements of smallpox.)

In any case, in 1796, a Scotch physician, Edward Jenner, tested this theory by infecting a boy with material from a pock on the hand of a milkmaid who had cowpox. He continued such experiments and it gradually became clear that people so infected were, indeed, resistant to smallpox.

The fancier medical term for cowpox is *vaccinia*, from the Latin "vacca," meaning "cow." A preparation containing material from a cowpock is a *vaccine* and the deliberate infection of a person with it is *vaccination.*

The term, however, has passed far beyond its original connection with cows. A generation later, the French chemist Louis Pasteur used the word *vaccine* to describe his own preparations of various germs with which he hoped to make human beings resistant to certain diseases, preparations that had nothing to do with cows. The best-known preparation now is one developed by Dr. Jonas Salk containing dead or nearly dead infantile paralysis virus — the so-called *Salk vaccine.*

Vector

THERE ARE two broad classes of measurements in science. One is simply a "how much" measurement or a "how many." You might say that there are two apples in a basket or that a line is 5.476 inches long or that an elephant weighs 2732 pounds or that there are 60 minutes in an hour or that an angle is 45 degrees. These are scalar quantities.

The word *scalar* comes from the Latin "scala," meaning a "ladder" or "staircase"; something, that is, with a succession of rungs or steps that can be counted off. That, after all, is all you are doing — counting off. It may be units or inches or pounds or minutes or degrees, but it is only a matter of counting. A device for measuring one of the most common of the scalar quantities, weight, is called simply a *scale*.

Sometimes counting isn't enough. You have to ask not only "how much" but "in what direction." You may exert a push of two pounds, for instance, but it is important to ask where you're pushing. In one direction, you may be guiding a girl across a dance floor; in another direction, you may fall out of a window. Forces, in other words, are not scalar quantities, but are vector quantities.

The word *vector* comes from the Latin "vehere," meaning "to carry" (the past participle of the Latin word is "vectus"). The notion of carrying comes from the fact that in any vector quantity there is the implication of something being carried from here to there. Thus, my two-pound push carries an impulse from here (my body) to there (my dancing partner's body, or through the window glass, as the case may be).

Bacteriologists use the word in a sense that has a more direct connection with its derivation. Some diseases are carried from person to person (or from animal to person) by some intermediate creature. Some species of mosquito, for instance, may carry malaria from one person to another. Malaria is therefore said to have an insect vector.

Velocity

THE MOST common word for the rate at which a body moves — that is, the distance covered by it in a given time — is *speed*. This comes from an Anglo-Saxon word meaning "success" and is still used in this sense when we "speed the parting guest." We are not trying to get rid of him quickly, of course; we merely wish him success on his way. Presumably, there was the general feeling that anyone who went about things quickly was more likely to succeed than another who was sluggish, so speed got the meaning of rate of movement and began to imply quickness rather than greatness.

In ordinary speech, *velocity* (from the Latin "velox," meaning "swift" and "velocitas," "swiftness") is used as a synonym for speed. If there is any difference, it is that *speed* is usually applied to the motion of a living thing, as a race horse, and *velocity* to an inanimate object, as a bullet.

In physics, however, there is an important difference. *Speed* is simply rate of movement. A car may have a speed of twenty miles an hour. *Velocity*, however, implies not only rate of movement but direction of movement. Thus, a car may have a velocity of twenty miles due north. Two cars moving in opposite directions may have the same speed but must have different velocities.

A moving object may change its velocity as time goes by. It may slow down or speed up or change its direction of motion. The rate at which velocity is changing is *acceleration*, from the Latin "ad-" (to) and "celare" (to hasten). Acceleration, in other words, means "to hasten toward."

Strictly speaking, acceleration refers to any change in rate of motion, but in popular speech, the implication of the derivation holds and acceleration applies only to motion that is speeding up. A slowing of motion is referred to as *deceleration*. The Latin prefix "de-" can mean "from" so that deceleration means taking "quickness from" motion.

Rocket experiments these days have introduced new phrases. "Positive acceleration" refers to either acceleration or deceleration that causes blood to move toward the head; "negative acceleration" to that which causes it to move toward the feet.

Ventricle

THE LATIN word for "belly" is "venter" and since the belly is a hollow place within the body, it came to be applied to other hollows as well. A small hollow would be "ventriculus" or, in English, a *ventricle.*

The most important ventricles in the body are the two main chambers or hollows into which the heart is divided: a left and right ventricle. Above each ventricle is a smaller chamber into which the blood enters on the way to the ventricle. It is the antechamber of the ventricle, so to speak, and is often called the *atrium,* which is Latin for "antechamber." However, the point where the blood enters each atrium bulges up like a small ear. The Latin word for ear is "auricula" and that word has come to be applied to the entire atrium. The two small chambers are commonly called left and right *auricle.*

The auricle contracts and forces blood into the ventricle, which contracts in its turn and forces the blood out of the heart altogether into an artery (see ARTERY). When the ventricle contracts, blood must not pass back into the auricle and to prevent that there is a *valve* (from the Latin "valvae," meaning "folding door") between each auricle and ventricle, which will allow blood-passage only one way, from auricle to ventricle, not vice versa.

The valve between left auricle and ventricle consists of two triangular flaps which look, when closed, like a bishop's miter (a two-pronged conical hat) and is therefore called the *mitral valve,* or *bicuspid valve,* from the Latin "bi-" (two) and "cuspis," genitive "cuspidis" (point). The valve between right auricle and ventricle is made up of three flaps and is the *tricuspid valve.* The Latin prefix "tri-" means "three."

When the left ventricle contracts, the blood is pushed into the aorta, the largest artery of the body. The aorta extends straight up at first, then loops and travels down the trunk. The first upward extension, however, makes the aorta look like a handle by which the heart is raised or held up and, indeed, the word *aorta* comes from the Greek "aeirein," meaning "to raise."

Vertebra

AN AXLE is the shaft on which a wheel, or wheels, turns. The Latin word for "axle" is *axis* and that is used specifically for the line (or imaginary shaft) about which a sphere turns, as, for instance, the axis of the rotating earth.

We have such lines of turning in our own body. The anatomist calls the armpit by its Latin name of *axilla*, which, like "ala," meaning "wing" is related to the word *axle*. Raise your arm stiffly and you will see that it turns about a line (or axle) running through the shoulder above the armpit.

A more direct example exists in the backbone. This is also called the *spine*, from the Latin word "spina" (thorn) because the individual bones of the spine are irregular in outline and have projections like thorns. These individual bones are *vertebrae* (singular, *vertebra*), from the Latin word "vertere" (to turn) because joints between particular vertebrae allow the turning of the head.

In man there are 33 vertebrae all together. The top 7 are the *cervical* vertebrae, from the Latin "cervix" (neck); then 12 *thoracic* vertebrae, from the Latin "thorax" (chest), also called *dorsal* vertebrae from the Latin "dorsum" (back); then 5 *lumbar* vertebrae, from the Latin "lumbus" (loin).

So far, the names indicate the position of the bones. But then, we have 5 *sacral* vertebrae, from the Latin "sacrum" (sacred) because this portion of an animal was used in sacrifices, and finally 4 *caudal* vertebrae, from the Latin "cauda" (tail). These last 4 vertebrae do actually represent the last traces of a human tail.

We come back to *axis* because the second cervical vertebra has that as its special name. It is the axis, because it is upon that particular vertebra that the head (plus the first vertebra) turns.

The four caudal vertebrae are sometimes lumped under the single word *coccyx*, from the Greek "kokkyx" (cuckoo) because the early anatomists thought the bone combination resembled the beak of a cuckoo.

Virus

When the French chemist Louis Pasteur was studying the dreaded disease hydrophobia in the 1880's, he noticed that the spinal tissues of affected animals could transmit the disease; yet he could detect no bacteria in the tissues. Pasteur had invented the germ theory of disease, and this observation did not shake him in his belief that diseases were carried by microscopic organisms. In this case, he speculated, the organisms were so small that they could not be seen by the microscope, that was all. And as it turned out, he was right.

In 1892, the Russian investigator D. Ivanovski made a mash of leaves from tobacco plants suffering from "mosaic disease" (in which their leaves were mottled and discolored) and passed the liquid from the mash through filters so fine that even bacteria could not pass through. The bacteria-free liquid that emerged could, however, still pass the disease on to healthy tobacco plants.

As was true of many Russian discoveries, this one was ignored by the West until one of their own scientists had repeated it. In this case, it was the Dutch investigator, M. W. Beijerinck.

At first, this was puzzling. An infective liquid free of living agents of disease? The word "virus" in Latin means "poison" (as in *virulent*) and this name was often applied to something that seemed to carry disease. Now the infective liquid was called a *filtrable virus* ("poison that passed through a filter"). In 1916, however, the American H. A. Allard used a finer filter than had any of his predecessors and held back the virus. The liquid that came through was not infective. So the virus was still a living agent except that, as Pasteur had guessed, it was a particularly small one.

Finally, in 1935, the American biochemist Wendell M. Stanley separated out pure tobacco mosaic virus (Ivanovski's virus), crystallized it and showed it to be a complex protein molecule. Such molecules still retain the old mystery in name, at least. They are *virus molecules*.

Vitamin

By 1900, it had become obvious that some diseases were connected with diet. For over a century, scurvy had been prevented by the use of citrus juices (see ASCORBIC ACID). Similarly, after 1878, the Japanese navy cut down on the incidence of the serious nerve disease beri-beri by forcing their sailors to eat brown rice rather than polished white rice. The Russian biochemist Nikolai Ivanovich Lunin showed in 1881 that rats died on a diet of purified carbohydrates, fats, and proteins, but survived if a small quantity of milk were added.

In 1901, the Dutch biochemist Gerrit Grijns first suggested that diseases such as scurvy and beri-beri were caused by the lack of some particular chemical in the diet, needed in very small quantities, and supplied by foods such as milk, fruit juices, rice husks, and so on. Because these diseases are caused by a dietary deficiency, they are called *deficiency diseases.*

The chemical in rice husks that prevented beri-beri proved to be an amine (see AMMONIA), and the Polish-born American biochemist Casimir Funk, supposing that the whole group of such chemicals might be amines, proposed, in 1912, the name *vitamine* for the group, from the Latin "vita" meaning "life." They were "amines of life."

It was soon found out however that most of the chemicals in question were not amines. The *e* was therefore dropped to reduce the effectiveness of the allusion and the word is now simply *vitamin.*

It early became obvious that there were at least two vitamins: one soluble in fat, not in water; the other soluble in water, not in fat. The American biochemist Elmer V. McCollum, in 1915, called these *fat-soluble A* and *water-soluble B.* By 1920, when the word *vitamin* was well established, the names *vitamin A* and *vitamin B* became common. In the years that followed, vitamin B was shown to be a mixture of many substances which became known as vitamin B_1, vitamin B_2, and so on, up to vitamin B_{12} and even beyond.

Vitriol

In ancient times, *transparent* objects (from the Latin "trans," meaning "across," and "parere," meaning "to appear"; light can, in other words, travel across a transparent object and appear on the other side) were unusual and names were derived from the fact (see CRYSTAL).

The Latin word for "glass," for instance, is "vitrum," so things that are glassy in appearance are *vitreous*. The jellylike transparent fluid inside the eyeball is called *vitreous humor*, for instance. (*Humor* means "fluid," in this case; see HUMOR.) The more watery liquid in front of the eye's lens is the *aqueous humor*, from the Latin "aqua" (water).

In the early Middle Ages, vitreous or glasslike minerals came to be called *vitriols*. The first mineral to be named so (in about 600 A.D.) was iron sulfate which was called *vitriol of Mars*, Mars being the alchemical name for iron (see MERCURY).

In time, a number of *sulfates* (minerals with molecules containing a sulfur atom and four oxygen atoms combined with any of various metal atoms) received the name, each being distinguished by color. Copper sulfate, for instance, forms beautiful blue, semitransparent crystals, and is called *blue vitriol* (or simply *bluestone*); while zinc sulfate forms colorless crystals and is *white vitriol*. The original vitriol, iron sulfate, forms green crystals and is *green vitriol*.

If vitriols are heated strongly, vapors are given off. About 1200 it was discovered that if these vapors are trapped, cooled, and dissolved in water, an oily and very corrosive liquid is formed. It was called *oil of vitriol*, or sometimes simply *vitriol*, though its proper name today is sulfuric acid.

The term *vitriolic*, as applied nowadays to a caustic and cutting remark or to a sharp-tongued and bitter personality, refers to the strong and caustic sulfuric acid and not to the innocent glass from which all these names are derived.

Volcano

The god of fire in ancient times was Hephaestus to the Greeks and Vulcanus to the Romans. The ancients associated fire chiefly with metals, since it was necessary, first, in the smelting of ore to obtain metal and then in the melting or softening of the metal so that it could be shaped and molded. It was natural, then, to think of the god of fire as being a divine and wonder-working smith. Hephaestus is pictured in just this way in Homer's *Iliad.*

At the same time, the ancients could not help noticing Mt. Etna in Sicily, a phenomenon among mountains. There were horrible noises within it, smoke issued from it, occasionally fire and molten rock belched out of it. Obviously it contained a gigantic forge, and later Latin poets made Etna the workshop of Vulcanus.

In Italian, the name Vulcanus has become Vulcano or Volcano (it is simply Vulcan, in English) and the word *volcano* came to be applied first to Mt. Etna and then to any mountain that behaved like Etna.

The opening through which the fiery phenomena make their appearance is the *crater.* This is from the Greek "krater," meaning a large bowl for mixing water and wine, which is actually what the volcano opening resembles during its quiet interludes. The hollowed mountains on the moon are called craters, too, but this is a poor name because it implies volcanic action whereas actually, those craters were probably formed by meteor collisions.

There is another famous volcano, Mt. Vesuvius, on the Italian mainland near Naples, and the Neapolitans have given us a well-known word in connection with volcanoes. The Italian word for "to wash" is "lavare." A rainstorm so violent as to flood the streets and, so to speak, wash them under a current of water, was called a "lava" by the Neapolitans. The word was naturally applied to another much more horrible sort of stream supplied occasionally by the neighboring Vesuvius, so that today *lava* means the stream of molten rock that pours out of an erupting volcano.

Volt

The units of electricity are a sort of Hall of Fame for various pioneers in the field (see power). In 1800, for instance, the Italian scientist Allesandro Volta produced electricity by piling up disks of silver and zinc alternately with brine-soaked paper between each silver-zinc pair (or "cell"). This was called a *Voltaic pile.* (Any group of similar objects is a "battery of objects." The Voltaic pile was a battery of electric cells and, eventually, simply a *battery.*)

The *electromotive force* (the Latin word "motivus" means "causing to move" — hence the "force that causes electricity to move") increases with the number of cells in the battery. This force is measured in *volts* in honor of Volta.

The current strength (that is, the quantity of electricity moving through a conductor each second) is expressed in *amperes.* This is in honor of André Marie Ampère, a French physicist, who from 1820 on, investigated the relationship of electricity and magnetism.

The total quantity of current moving through a conductor over a period of time is measured in *coulombs.* This commemorates the French physicist Charles Augustin Coulomb who, from 1785 on, worked out the manner in which electric charges attract and repel each other.

In 1827, the German physicist Georg Simon Ohm published a pamphlet putting forth *Ohm's Law.* The law states that when electric current flows through a conductor, the electromotive force divided by the current strength gives a figure which represents the resistance of the conductor. This pamphlet made no impression at the time and the sensitive Ohm resigned his professorship at Cologne in mortification. However, the unit of electrical resistivity is today called the *ohm* in his honor. Moreover, electrical conductivity (one divided by the resistivity) of any conductor is the *mho,* which is only *ohm* spelled backward. Thus, a wire with a 5-ohm resistivity has a ⅕-mho conductivity.

Vulcanize

When Christopher Columbus landed on the coast of South America, he found the natives bouncing balls of an elastic substance and playing games with them. Nothing like it was known in Europe. The natives obtained this material from a milky liquid that oozed out of a certain tree when the bark was cut. The natives called the substance "cahuchu," meaning "weeping wood." (We have the more prosaic name of *latex* for the fluid, which is nothing more than the Latin word for "fluid.")

The French call the bouncing substance *caoutchouc* and the Spaniards *cauch* from the original Indian word; but the English is far less romantic. The English chemist Joseph Priestley noted that the substance could be used to rub out pencil marks, so about 1770 he called it *rubber* and the name stuck. The English often call it *India rubber* because it comes from the "Indies," the old-fashioned name for the Americas, dating back to Columbus's belief that he had landed in India.

One of the first uses of rubber was as a waterproofing material. The Englishman Charles Mackintosh, for instance, first prepared rubber-coated cloth, and raincoats are still sometimes called *mackintoshes* for that reason.

However, there was this difficulty. Rubber got stiff and brittle in cold weather and soft and sticky in warm. The American Charles Goodyear was one of those who tried to correct this failing. He succeeded by accident. In 1839, he was mixing rubber and sulfur to see if that would help, and accidentally he tipped some over onto a hot stove. He got it off as quickly as he could (what a stink it must have made!) and when he did so he found he had a piece of rubber that stayed elastic in the cold and dry in the heat. Because Vulcan was the Latin god of fire (see VOLCANO), the rubber and sulfur, united successfully under the influence of fire, was said to be *vulcanized rubber*. The success of modern rubber depends on this one discovery, but Goodyear spent his life fighting for patent rights and died loaded down with debts.

X Rays

STREAMS OF electrons were first produced in evacuated tubes in the 1860's, and for thirty years, physicists were fascinated by these *cathode rays* (so called because they emerged from the cathode; see ELECTROLYSIS) without really understanding them. In that period, the discoveries that were made seemed all the more mysterious because of that lack of understanding.

In 1895, for instance, the German physicist Wilhelm K. Röntgen noticed that a certain chemically coated paper in his laboratory glowed whenever his cathode-ray tube was in operation, even when there was cardboard between tube and paper. His cathode-ray tube would fog photographic plates, too, even when the plates were protected by their wrappings.

He decided that some kind of radiation was formed in the cathode-ray tube that could pass right through glass, cardboard, and paper. He had no notion as to what this radiation might be, and he called them *X rays.* This name suited the mystery surrounding the cathode rays, since X is the letter usually used by the mathematician to signify the unknown.

Nowadays we know that the cathode rays are really streams of electrons, and we know that X rays are quite similar to ordinary light except that they are much more energetic. Despite the greater understanding of today, Röntgen's radiation is still the "X ray." Sometimes, X rays are called *Röntgen rays* after the discoverer, but the name is difficult to pronounce, except for Germans, and the habit has not caught on.

The same happened to another type of radiation similar to light, but one that is much weaker — the so-called radio waves. The word *radio* is simply a general term that would fit any radiation. Here, too, there is an alternative name, *Hertzian waves,* after the discoverer, the German physicist Heinrich Hertz. And here, too, the meaningless name won out, and *radio waves* is the term used by practically everybody.

Yttrium

THE EARLY chemists used the name "earth" for any substance that did not dissolve in water and was not affected by heat. The five most common "earths" were silica, alumina, lime, magnesia, and iron oxide. Together, these make up 90 per cent of the earth's crust so the word "earth" is fitting. Lime and magnesia can be brought into solution by treatment with chemicals and such solutions had alkaline (see POTASSIUM) properties. Lime and magnesia were therefore called the *alkaline earths*. When, eventually, calcium and magnesium were found to occur in those earths, they and certain related elements were termed the *alkaline earth elements*.

In 1794, a Finnish mineralogist, Johan Gadolin, investigated a new black mineral found seven years earlier in a quarry in Ytterby, a small village near Stockholm, Sweden. Gadolin decided that the mineral contained a new kind of "earth," and reported this. The new "earth" was isolated and named *yttria* after Ytterby. (It has since been renamed *gadolinite*.)

It did not take long to discover that there were a number of different "earths" in yttria and other similar minerals. To distinguish them from the very common "earths" that were already known, these new ones were called the *rare earths*, and when new elements were discovered in them these were called the *rare earth elements*.

By 1843, the Swedish mineralogist Carl Gustav Mosander had divided yttria into three earths. For one he kept the name yttria, while the other two he called *erbia* and *terbia*, still after Ytterby. Then when, in 1878, the Swiss chemist Jean Charles de Marignac discovered yet a fourth "earth" in what Mosander had called erbia, he named the fourth "earth" *ytterbia*, again after Ytterby. Eventually, new metallic elements were discovered in each of these "earths" and they were named *yttrium*, *erbium*, *terbium*, and *ytterbium*, so that the insignificant hamlet of Ytterby ended with four elements named in its honor.

Zero

THE ANCIENTS used letters of the alphabet to represent numbers. The most familiar example today is the Roman system, still used on inscriptions, clock faces, and the like. In the Roman system, I is one, V is five, X is ten and so on.

Such a system did not involve positional values. In other words, it didn't matter where the X was in a number, it always stood for ten. Thus, XXX was ten plus ten plus ten, or thirty.

In the Middle Ages, a new system reached western Europe from India via the Arabs. In this system each number had a separate symbol ("Arabic numbers") which were not letters of the alphabet and which had positional value. For instance, 555 is not five plus five plus five, as it would be in the Roman system. The 5 at the right is five indeed, but the 5 in the middle stands for fifty while the 5 at the left stands for five hundred. The number is therefore five plus fifty plus five hundred, or five hundred and fifty-five.

This new system is so superior to the old that it is amazing the clever Greeks never thought of it. Apparently, the catch was that they never thought of using a symbol for nothing at all.

For instance, how would you distinguish between fifty-five and five thousand and five? On the abacus (see CALCIUM), the two numbers look somewhat the same. For fifty-five, the two bottom rows of counters have 5 pebbles pushed to the right. For five thousand and five, also, two rows have 5 pebbles pushed to the right; but between the two rows are two untouched rows. The Hindus invented a symbol for such an empty or untouched row (a thing the Greeks never did) and the Arabs adopted it and called that symbol "sifr" (empty).

This has come down to us as *cipher* or, in more distorted form, *zero.* Now we write fifty-five as 55, while five thousand and five is 5005. The importance of zero to mathematics is shown by the fact that *cipher* has also come to mean "to solve arithmetical problems."

Zodiac

It is a natural thing in observing the stars to run them together into patterns that resemble familiar things. The best example to most of us is the Big Dipper. Ancient peoples did this quite elaborately. The Greeks, for instance, divided all the visible stars into groups which we now call *constellations*, from the Latin prefix "con-" (together) and "stella" (star). Constellations are stars grouped together, in other words.

In general, the ancients were quite imaginative in what they saw in the constellations. They saw bears, dogs, winged horses, serpents, crows, and assorted people.

Certain constellations were particularly important. The sun, the moon, and the five visible planets, as they moved against the background of the stars, restricted their motion to a relatively narrow band in the sky, passing through only certain constellations. This band is called the *ecliptic*, (Greek, "ekleipsis") because in it eclipses of the sun and moon take place. The word *eclipse* (referring to times when the moon passes into the earth's shadow, or vice versa) comes from the Greek words "ek" (out) and "leipein" (leave). During an eclipse, the sun or moon is "left out" of the sky, so to speak.

The stars of the ecliptic are divided into twelve constellations. The moon travels through all twelve in 27 days; the sun spends one month in each; the planets pass through the cycle in more complicated patterns. The constellations are Aries (the Ram), Taurus (the Bull), Gemini (the Twins), Cancer (the Crab), Leo (the Lion), Virgo (the Maiden), Libra (the Scales), Scorpio (the Scorpion), Sagittarius (the Archer), Capricornus (the He-Goat), Aquarius (the Water-Bearer) and Pisces (the Fishes).

Seven of the twelve constellations are animals. The Greek word for "animal" is "zoon." A diminutive of "zoon" is "zodion" and the adjective from "zodion" is "zodiakos." In talking of the constellations of the ecliptic, you are talking mostly of animals, and they are known as the *zodiac* to this day.

Index

Pleistocene Rays
stice Yttrium Rad
sium Quantum
Almanac Ascorbic
ower Spectrum
ermat Sulfanilam
amite Ammonia
Iodine Alpha A
ucose Meridian Rh
sec Thermonuclear A
e Zero Amalgam
igonometry React
n Volcano Irratio
niverse Zodiac Infra
ium Uranium Spira
s Argon
Rhinoce
tercl Ungulate Calci